Law Relating to Biotechnology

Law Relating to Biotechnology

Sreenivasulu N.S.

OXFORD
UNIVERSITY PRESS

OXFORD
UNIVERSITY PRESS

Oxford University Press is a department of the University of Oxford.
It furthers the University's objective of excellence in research, scholarship,
and education by publishing worldwide. Oxford is a registered trademark of
Oxford University Press in the UK and in certain other countries

Published in India by
Oxford University Press
YMCA Library Building, 1 Jai Singh Road, New Delhi 110 001, India

ISBN-13: 978-0-19-946748-8
ISBN-10: 0-19-946748-X

Typeset in Adobe Jenson Pro 10.7/13.3
by The Graphics Solution, New Delhi 110 092
Printed in India by Replika Press Pvt. Ltd

Contents

Preface

Biotechnology probably is the most promising and, at the same time, most controversial technology that the modern world has ever seen. Though it has opened varied and abundant opportunities for prosperity, it has also raised a number of issues and challenges. Yet the human society is poised to nurture and use biotechnology in the most viable and balanced way without compromising on the sustenance, forbearance, and growth of nature and ecology. Given its diverse commercial and industrial applications, biotechnology promises a future that caters to different needs of the community at large.

There are a number of biotechnology-related legal challenges that need to be addressed by regulating its promotion, sustenance, use, and commercial utilization under a legal governance system. Biotechnology has developed from the classical age to the modern age, adding to its strength, acumen, and efficiency in fulfilling the needs of the modern society. A range of techniques and methods developed with the help of biotechnology have empowered human beings in not only challenging nature but finding solutions to problems that were once thought to be unresolvable. Its commercial application and industrial viability make it a much sought-after technology. All these traits have made it imperative for the governing systems to mull over the kind of policy to be developed and adopted for regulating biotechnology. The international society has also responded well to this great technology and effectively used it for sustainable development. Legal regimes have been altered, reformed, and developed to promote the use of technology toward commercialization. Intellectual property issues related to the protection of innovations of biotechnology have been dealt effectively under the settled intellectual property law.

The trade-related aspects of biotechnology have been addressed under the international trade law read with the regional and domestic underpinnings. The environmental concerns involved in promoting biotechnology have been specifically addressed while working toward sustainable development. Attempts have been made to balance the biodiversity aspects of biotechnology in the light of debates between bioprospecting and biopiracy. At the same time, international society is responding to all the human rights challenges related to biotechnological innovations in a scientific manner. Nations are promoting biotechnology at the domestic levels quite enthusiastically in the light of the international societies' guidance and support. Nations such as the US and the UK are exploiting the commercial uses of biotechnology while effectively bringing it under regular legal regimes. Whereas, nations like India have been cautious and progressing gradually in regulating biotechnology at the required levels.

At this juncture, the current work attempts to analyze the development of biotechnology through different ages and generations toward commercial application in the modern world while emphasizing upon its varied industrial applications. It throws light on the various policy concerns related to biotechnology at the international, regional, and domestic levels. A number of legal issues pertinent to biotechnology, such as intellectual property issues, trade-related aspects, environment and biodiversity concerns, and human rights underpinnings, have been discussed, debated, and analyzed in a scientific yet legal perspective. The Indian legal regime on biotechnology regulation is also showcased toward the end. I believe that the work would be useful for professionals, academia, researchers, and students working in the fields of biotechnology, law, and life sciences. At the same time, it would be quite handy for those who would like to understand the interface between biotechnology, law, and policy toward a logical conclusion.

Abbreviations

ADA	adenosine deaminase
CEC	Codex Alimentarius Commission
COP	Conference of the Parties
CBD	Convention on Biological Diversity
DBT	Department of Biotechnology
EC	European Commission
ECJ	European Court of Justice
EFB	European Federation of Biotechnology
EPO	European Patent Office
EU	European Union
FAO	Food and Agriculture Organization
GMO	genetically modified organism
GEAC	Genetic Engineering Approval Committee
GE	genetically engineered
HGS	Human Genome Sciences

ICMR	Indian Council of Medical Research
IBSC	Institutional Biosafety Committee
IPR	Intellectual Property Rights
ICESCR	International Covenant on Economic, Social, and Cultural Rights
LMO	living modified organisms
GATT	General Agreement on Trade and Tariff
PCR	Polymerase Chain Reaction
PBR	Plant Breeder Right
RCGM	Review Committee on Genetic Manipulation
rDNA	recombinant DNA
RDAC	Recombinant DNA Advisory Committee
SOP	standard operating procedure
SBIR	Small Business Innovation Research Initiative
TEC	Technical Expert Committee
TRIPS	Trade-Related Aspects of Intellectual Property Rights
UDHR	Universal Declaration of Human Rights
UNESCO	United Nations Educational, Scientific, and Cultural Organization
UDHGHR	Universal Declaration on the Human Genome and Human Rights
USPTO	United States Patent and Trademark Office
UPOV	Union for the Protection of New Varieties of Plants
WIPO	World Intellectual Property Organization
WTO	World Trade Organization

1 Biotechnology

The Science of Wonders

"Life is wonderful, make it colorful" is not a hypothesis of a poet advocating for a cheerful life; it is an executable proposition. "Science is wonderful and it makes life colorful" could in fact be the catchline of the modern world. Science has added color to human life, and progress in science has opened up our lives to previously unimagined possibilities. In today's times, scientific innovations and new technologies are coming up at a fast pace. Humankind is touching new heights and witnessing new merits in the sphere of technological advancement. The introduction of new technologies shows the thirst of humankind for development and the zeal to know new things. Science and technology have profoundly influenced the course of human civilization. The scientific revolution of the twentieth century led to many new technologies, which promised to herald new eras in many fields. The twenty-first century seems to be shaped in a significant measure by the development of biotechnology,[1] which could also be called the exploitation of living organisms for the industrial utilization.[2]

As we stand today, well into the twenty-first century, we have to ensure optimal use of these developments for the well-being of our people. Science and technology have been an integral part of the Indian civilization and culture over the past several millennia. Few are aware that India was the fountainhead of important foundational scientific developments and approaches. These cover many great scientific discoveries and technological achievements in mathematics, astronomy, architecture, chemistry, metallurgy, medicine, natural philosophy, and other areas. A great deal of these discoveries traveled to different countries from India. Simultaneously, India also assimilated scientific ideas and techniques from other countries, with open-mindedness and a rational attitude characteristic of scientific ethos. The principles of universal harmony, respect for all creation, and an integrated holistic approach form the foundation of Indian traditions. This background would provide valuable insights for future scientific advancements. During the century prior to Independence, there was an awakening of modern science in India through the efforts of a number of outstanding scientists, who were responsible for great scientific advances of the highest international caliber.[3]

Technology is an application of knowledge to solve specific problems or meet identified needs. It is a result of both physical and mental human endeavors. Furtherance of knowledge can be attributed to the development of technology. Development is an unending process and progress is a journey with infinite destinations. The role of technology is significant in the continual process of development and progress. It is a result of progress and development toward achieving specific goals. Since the evolution of civilization, we have seen many developments, witnessing the progress of the society.[4] Technology encompasses

---

[1] Naidu, D. 2009. *Biotechnology and Nanotechnology: Regulation under Environmental, Health and Safety Laws*, p. 1. London: Oxford University Press.

[2] Sreenivasulu, N.S. 2013. *Law Relating to Intellectual Property*, 1st edn, p. 344. IN, USA: Penguin-Partridge Publications.

[3] Council for Scientific and Industrial Research. 2003. 'Science and Technology Policy of India', available at http://www.csir.res.in/external/heads/aboutcsir/Policy2003.htm (last accessed 3 February 2016).

[4] Sreenivasulu, N.S. 2008. *Biotechnology and Patent Law: Patenting Living Beings*, 1st edn, p. 1. Noida: Manupatra Publications.

the application of scientific principles to commercial uses. Technology refers to innovations in almost any field, from the life sciences such as medicine (for example, development of a new drug and medical device) or gene therapy (biotechnology), to manufacturing (industrial science), to the more abstract sciences such as aerospace engineering. Recently, however, it has come to mean the creation, transmission, and storage of data and information using hardware (computers, mobile phones, or telephone lines) and software (computer programs).[5] Technological advancements are a reflection of the progress of the society. Since the ancient times, different technologies have marked the progress of the society. Starting from the invention of fire, cultivation, animal husbandry, motor, fermentation technology, steam engine, printing machine, telephone, aviation, radio, television, computer technology to information technology, modern society has made significant progress. Modern world can be seen as a hub of many technologies that have made human life a manifested world of execution of human thoughts. Baroness Susan Greenfield mentioned that the twenty-first century is facing challenges that are unique in itself, while speaking in the House of Lords in April 2006. This is because science is now delivering a diverse range of information technology, nanotechnology, and biotechnology, with a speed and convergence that we could never have predicted even a decade ago.[6] Science always brings something special to humankind. Scientific progress and technological advancements have always resulted in the betterment of the society. According to a paper presented and discussed at the President's Council on Bioethics, we have entered into a golden age of biology, medicine, and biotechnology. With the completion of the Human Genome Project and the emergence of stem cell research, we can look forward to major insights into human development, both normal and abnormal. The once-discrete medical disciplines of anatomy, physiology, biochemistry, bacteriology, pharmacology, and pathology have coalesced into a single, powerful flow of knowledge that can be expressed in the universal language of chemistry. Similarly, the emerging biological disciplines of genetics, cell biology, developmental biology, and

---

[5] Aspatore. 2004. *Inside the Minds: The Laws Behind Technology: Leading Lawyers on the Legal Aspects of Patents, Software Licensing, Telecommunications & More*, p. 1. USA: Aspatore Books.

[6] Brownsword, Roger. 2008. *Rights, Regulation, and the Technological Revolution*, p. 1. New York: Oxford University Press.

molecular biology come under the umbrella of biotechnology. At times, forces obstruct the flow, creating eddies and diverting it from its main course, by which it approaches ever closer to the secrets of nature. The same kind of flow and force has enabled the emergence of medical and biological sciences in this century.[7] Technological advancement in life sciences has resulted in a new technology called biotechnology, also known as the field of modern-day miracles.[8] The advent of biotechnology has revolutionized the progress in science. However, describing a technology like biotechnology involves mental and intellectual endeavors.

It is said that, mythologically speaking, our fate is in stars, but biotechnologically speaking, our fate is not in stars, it is in our genes. Biotechnological breakthroughs and the innovations of genetic engineering have not only altered the assumptions of mythology, but created and coined new and novel assumptions based not on beliefs but on scientific experiments. Biotechnology has influenced and changed the modern world in an innovative and scientific way. Different technological revolutions had a great impact on our lives. Scientific advancements and technological developments have many a times given rise to dramatic changes in the society. These advancements and developments have not only been milestones in the history of humankind in terms of refining the human life, but also achievements that have strengthened and empowered us in terms of refining and redefining the way we live in the society. Similarly, biotechnology is poised to empower the humankind to redefine the life in the contemporary world. Since the science of biotechnology is showing enormous potential in changing not only the way we think but also how we live and perceive life, it is considered as the science of modern-day miracles. The wide-ranging scope of biotechnology has resulted in its rediscovering life on the earth; it promises to have the potential to make dreams come true by making miracles possible.[9] It continually

[7] Kornberg, Arthur. 1995. *The Golden Helix: Inside Biotech Ventures*, p. 4. CA, USA: University Science Books.

[8] See Sreenivasulu, N.S. 2005. 'Biotechnology and Patent Law,' *Journal on World Intellectual Property Rights*, 1–2 (July–December): 1.

[9] See Sreenivasulu, N.S. 2007. 'New Face of Patent Law: Impact of Information Technology and Biotechnology Revolutions,' *Manupatra Intellectual Property Reports*, 1(3): 135.

mesmerizes the ordinary person by assuring remedy for many odds that were once considered irremediable.

Defining Biotechnology

With biotechnology being a dynamic discipline where exponential progress is being made, it is difficult to define it in precise terms. Despite the attempts that are being made to define biotechnology, there is no universally accepted definition.[10] While there are multiple definitions of biotechnology, yet there is great confusion regarding its theoretical meaning and practical application.[11] Many organizations and working parties have published reports that attempt to define biotechnology.[12] The United Nations Congress Office of Technology Assessment[13] defines biotechnology to mean "any technique that uses living organisms to make or modify a product to improve plants or animals or to develop microorganisms for specific uses." The European Federation of Biotechnology (EFB) defines biotechnology as the integration of natural sciences and organisms, cells, parts thereof, and molecular analogs for products and services.[14] British biotechnologists, like Smith, define biotechnology as an "application of biological organisms system or processes to manufacturing and service industries."[15] Japanese biotechnologists define biotechnology to mean a technology using biological phenomena for copying and manufacturing various kinds of useful substances.

[10] Parsley, G.J. 1990. *Agricultural Biotechnology: Opportunities for International Development*. Wallingford, UK: C.A.B. International.

[11] Ranga, M.M. 1999–2000. *Animal Biotechnology*, p. 2. Jodhpur: Agro Bios India.

[12] Sreenivasulu, N.S. 2006. 'Information Technology and Biotechnology', in T. Sabanna (ed.), *Globalisation and WTO*. New Delhi: Serials Publications.

[13] Office of Technology Assessment. 1981. 'United States Office of Technology Assessment Report', Washington DC, available at http://ota.fas.org/reports/8105.pdf (last accessed 2 February 2016); Office of Technology Assessment. 1984. 'United States Office of Technology Assessment Report', Washington DC, available at http://ota.fas.org/reports/8400.pdf (last accessed 2 February 2016).

[14] Smith, J.E. 1996. *Biotechnology*, 3rd edn, p. 2. Cambridge, UK: Cambridge University Press.

[15] Smith. 1996. *Biotechnology*.

However, biotechnology is broadly defined as any technique that uses living organisms or parts of organisms to make or modify products to improve plants or animals, or to develop microorganisms for specific industrial purposes.[16] It is not a single technique, rather a combination of different techniques utilized to manipulate living organisms in directed fashions. In simple sense, biotechnology means the application of technology or scientific knowledge to biological processes.[17] Accordingly, biotechnological invention is defined as "any invention which concern a product consisting of or containing biological material or process by means of which biological material is produced, processed or used."[18]

Understanding Biotechnology

Biotechnology is a study related to the practical application of living beings in different fields for industrial purposes.[19] Different techniques are combined to utilize for manipulation of living organisms in directed fashions. It involves human interventions in the natural biological processes by way of technical contribution to arrive at desired results. Therefore, biotechnology could be understood as an application of human-invented techniques to natural biological processes. The study of genetics is often referred to as biotechnology; however, genetics is a subdivision of biotechnology, which in fact is a combination of a number of techniques and methods used to modify the functioning of living beings and living processes. In fact, it is genetic engineering or research in genetics that attracts human rights concern more than any other technique of biotechnology. In some ways, there is nothing new about biotechnology, as the use of biological systems, living organisms, or derivatives thereof to make or modify products or processes for specific use had been happening since the earliest human civilizations.[20] For instance,

[16] Sreenivasulu. 2008. *Biotechnology and Patent Law*, pp. 2–3.

[17] Cannon, B.C. 1994. 'Toward a Clear Standard of Obviousness for Biotechnology Patents', *Cornell Law Review*, 79(3): 735–65.

[18] Chapter VI, Rule 23B(2) of the European Patent Convention (EPC).

[19] Sreenivasulu, N.S. 2011. *Intellectual Property Rights*, 2nd edn, p. 49. New Delhi: Regal Publications.

[20] Sreenivasulu, N.S. 2005. 'Patenting Genetically Modified Life Forms: Legal Issues and Challenges', *Indian Bar Review*, XXXII(3&4): 485–98.

biotechnology has helped to conserve food and produce beer and yoghurt, and to treat waste water since ages. However, in the 1970s, in vitro nucleic acid techniques, which allow fusion cells beyond the taxonomic family that overcome natural physiological reproductive or recombination barriers and techniques, were referred to as modern biotechnology. It opened up a new range of possibilities for studying and making use of biological mechanisms.[21] Since then, biotechnology has evolved rapidly, venturing into many new areas and leading to the development of a large number of applications in diverse industrial fields such as fermentation processes, medicine, diagnostics, pharmaceuticals, food production and food processing, plant agriculture, animal agriculture, energy and environment management, among others. Biotechnology is a combination of different techniques involving different disciplines of science. According to the scientific community, it is a combination of different technologies put to use simultaneously. It brings together basic sciences and advanced sciences in its realm—it combines life sciences and chemical sciences; biology, botany, zoology, chemistry, microbiology, biochemistry, genetics, agriculture, tissue culture, cell culture, fermentation technology, hybridoma technology, recombinant technology, protein engineering, process engineering, and genetic engineering. Here, biology covers a number of different studies comprising ecology, physiology, genetics anatomy, taxonomy, microbiology, and parasitology. Genetic engineering involves taking genes from one cell and inserting them into another cell, which gives the cell new characteristics that are determined by the transferred gene. It is concerned with how characteristics inherited from one generation pass onto the next generation. Technically speaking, biotechnology involves techniques to modify living beings for desired results. Scientifically speaking, it is a study related to the refinement of living organisms. Technologically speaking, it is a combination of different techniques to get better results. Historically speaking, it is an art of producing fermented foods, wine, beer, and cheese. Generally speaking, it is a combination of life and technology for an intended purpose. Broadly, it is a technique that involves modifying the living organisms such as microorganisms, plants, and animals to incorporate certain desired

---

[21]   Wuger D. and T. Cottier. 2008. *Genetic Engineering and the World Trade System: World Trade Forum*, pp. 3–4. New York, USA: Cambridge University Press.

nonnatural traits and qualities for enhanced life. It involves more than one technique; it is a collection of different techniques put to use together for manipulating and modifying living organisms in desired and directed approaches toward achieving intended goals. It involves using living organisms or parts of organisms to make or modify products, to improve plants or animals, or to develop microorganisms for specific uses. Simply speaking, biotechnology is a study of microorganisms and related things. It is a study related to the practical application of living beings in different fields.[22] It is a study related to living organisms in the industrial utilization.[23] It is the technology that uses living organisms or parts of living organisms for specific commercial use.[24] It intends modification and application of living beings for different practical purposes. It is about employing products and processes that are alive.[25]

Conceptualization

The history of mankind is interspersed by different revolutions. Society has been greatly influenced by different technologies that impacted the progress. Ranging from industrial revolution and electronic revolution to information technology revolution, humankind has faced and handled various revolutions remarkably. In the recent past, society has been influenced by yet another technological revolution called biotechnology, which is a culmination of human-invented technology and natural processes. Reports say that biotechnology is not new but has existed since ancient times. In those days, fermentation technology was used to produce and preserve goods for a long time.[26] With the addition of genetic engineering or recombinant technology, the same fermentation technology is now called biotechnology. In fact, it is genetic engineering that has

[22] Sreenivasulu. 2007. 'New Face of Patent Law', p. 140.

[23] Sreenivasulu. 2005. 'Biotechnology and Patent Law'

[24] Gabrielle, J. 1990. *Agricultural Biotechnology: Opportunities for International Development*, p. 31. Wallingford, UK: C.A.B. International.

[25] Arup, C. 2000. *The New World Trade Organization Agreements*, p. 224. UK: Cambridge University Press.

[26] See Sreenivasulu. 2005. 'Patenting Genetically Modified Life Forms', pp. 485–98.

brought this hidden technology to the limelight. Genetic engineering is capable of manipulating living organisms and make them perform and function in a way that goes beyond natural limitations. Advances made in genetic engineering has lead to biotechnology being labeled as the technology of modern-day wonders—production of genetically modified nonnatural living organisms being a possible reason for biotechnology being considered so.

The term biotechnology brings different thoughts to mind. Some think of developing new types of animals. Others dream of almost unlimited sources of human therapeutic drugs. Many envision the possibility of growing crops that are more nutritious and naturally pest-resistant to feed a rapidly growing world population. In its purest form, the term biotechnology refers to the use of living organisms or their products to modify human health and the human environment.

The term biotechnology is a combination of two words "bios," meaning life, and "technology," which is the application of scientific knowledge for industrial purposes. In essence, biotechnology is concerned with the application of scientific techniques to living organisms.[27] These living organisms are manipulated in one way or another to develop desirable traits for commercial purposes. Since its origin, biotechnology has been expanding its scope and significance. As biotechnology is a study related to the practical application of living beings in different fields,[28] so its scope knows no boundaries of any field. Biotechnology has become an integral part of modern-day life. It was considered an art when fermentation technology was practiced centuries back to prepare cheese, beer, and wine. It can be said that biotechnology became science from art with the value addition of a number of techniques and methods that were invariably multidisciplinary in nature and multifaceted in application. For thousands of years, humans have been manipulating the inherent characteristics of plants and animals through cross-hybridization and selective breeding to create desirable traits that either have commercial value or perform a useful function.[29] In the modern times, these

[27] Torremans, Paul. 2008. *Holyoak & Torreman's Intellectual Property Law*, p. 85. New York: Oxford University Press.

[28] Sreenivasulu. 2013. *Law Relating to Intellectual Property*.

[29] Naidu. 2009. *Biotechnology and Nanotechnology*, p. 5.

interventions aim at achieving intended result in the living being by either introducing a new gene and activating it or removing a gene or suppressing the activities of a gene. According to European Patent Office (EPO) guidelines, biotechnological inventions are those that concern a product consisting of or containing biological material or a process by means of which biological material is produced, processed, or used. The guidelines further state that biological material refers to any material containing genetic information and capable of reproducing itself or being reproduced in a biological system.[30] Biotechnology encompasses some very old technologies as well as some very new ones. For example, fermentation technology, the growing of microorganisms, has been practiced since the time human beings started making wine, beer, and bread. On the other hand, the techniques of genetic engineering, gene cloning, and gene splicing are the relatively new ones that also fall within the ambit of biotechnology.[31]

The Quest for Perfection and Urge for More

The pursuit of perfection and suitability has taken humankind to great heights. The quest for gaining ability and compatibility for deferent needs has yielded various benefits for the humankind. Humans have always been fascinated by the terms "perfect" and "more." There is no limit for more; there is nothing less than perfect to achieve more. Since the ancient times until the modern digital age, humans have always aimed for perfection in the pursuit of comfort, convenience, and happiness. That is why they have always looked for these things in various ways. Whether it is arts, culture, trade, commerce, profession, or sports, honors have always gone to the one who has reached perfection or who has achieved more. That is the reason "*Yeh dil mange more*" (this heart wants more), "believe in the best," "better than the best,"

[30] European Patent Office. 2015. 'Guidelines for Examinations in the European Patent Office', Part C, Chapter IV, Para 2a, Munich, available at http://www.epo.org/law-practice/legal-texts/html/guidelines/e/g_ii_5_1.htm (last accessed 2 February 2016).

[31] Epstein, M.A. 2008. *Epstein on Intellectual Property*, 5th edn, pp. 12–3. New Delhi: Wolters Kluwer (India) Pvt. Ltd.

"the perfect choice," and "competent and comprehensive" have become popular slogans in the contemporary world. In the pursuit of happiness, humankind has endeavored for more with perfection. It could be more resources, money, labor, and greater advancement, near perfection or just perfection to live the life with the most comfort levels or convenience. In the contemporary times, the terms "more," "better," "best," "perfect," and "competent" are used to describe who is what, what one can do, and what is his/her status. On similar lines, states with greater power, more money, advanced research, superior technology, and greater development are considered as developed and competent states that allow for more convenience, comfort, and better lifestyles. Among the states having the aforementioned features, the one closest to perfection is considered the best. The concepts "more" and "perfection" have been inducted into the minds of humans from the age-old theory: "the survival of the fittest." Applying the theory in the modern-day scenario, it is the survival of the strongest in terms of money, power, and technology. From the very beginning, this concept managed to hold water. From the prehistoric times until today, the stronger have prevailed over the weaker by their sheer power. Here physical power or arms are not the only measures of strength; it also includes technology. In the wars and battles that we have witnessed in the history of human civilization, the armies having better quantitative and qualitative manpower, superior arms, and better technology have won battles. The one who is the fittest, strongest, and has achieved greater perfection is better prepared for life. The race for compatibility, perfection, and a better way of life continues without any halt. The one who wins the race of competence and perfection becomes the master. This master race signifies the quest of human beings to get something that empowers them to lead the life in the best possible way. It is an urge for a system where living beings of one kind are superior to all the other kinds. This urge goes beyond the law of nature, which says that there is no perfect thing or being apart from God. However, humans have always challenged nature and endeavored to lead a better life. They have modified and changed many things by experimenting with them in their pursuit of perfection and of possessing more to lead lives that were more comfortable. In due course, human beings achieved many things that marked the emergence of different eras and ages.

Classification

Biotechnology has developed considerably and the field has witnessed progress across centuries. Based on origin and development, biotechnology can be classified into two different categories: classical biotechnology and modern biotechnology. The World Intellectual Property Organization (WIPO) study on the importance of Biotechnology for National Growth and Development[32] stresses on the significance of biotechnology in the modern world.[33] The European Federation of Biotechnology (EFB) considers biotechnology as the integration of natural sciences and organisms, cells, parts thereof, and molecular analogs for products and services. The EFB definition is applicable to both classical and modern biotechnology. Classical biotechnology refers to the conventional techniques that have been used for many centuries to produce beer, wine, cheese, and many other foods, while modern biotechnology embraces all methods of genetic modification by recombinant DNA (rDNA) and cell fusion techniques together with the modern developments of classical biotechnological processes. Let us understand classical and modern biotechnologies in detail.[34]

Classical Biotechnology

Biotechnology, defined as the application of scientific knowledge to transfer beneficial genetic traits from one species to another to enhance or protect an organism, has been part of the development of human culture. The earliest examples of biotechnology could be the domestication of plants and animals leading to a major change in lifestyle, from hunter-gatherers to settled farmers. Classical biotechnology comprises traditional or conventional methods and techniques mostly used and practiced in the period that started since the Sumerians brewed beer until the eighteenth century when modern techniques of biotechnology

---

[32] 'WIPO Study on the Protection of Inventions in the Field of Biotechnology'. 1987. European Intellectual Property Organization (EIPO) and Cornell University, New York.

[33] Gene Campaign. 1997. 'Study on the Importance of Biotechnology for National Growth and Development', New Delhi.

[34] Smith. 1996. *Biotechnology*, pp. 2–3.

were not invented. The production of ethanol from yeast cells and various industrial chemicals such as acetic acid and acetone by fermentation process was well known and being done since ages.[35] Prehistoric biotechnologists used yeast cells to make bread and ferment alcoholic beverages, and bacterial cells to make cheese and yogurt; they bred their strong, productive animals to make even stronger and more productive offsprings. Biotechnology is as old as the ancient culture of Indians, Chinese, Greeks, Romans, Egyptians, Sumerians, and other ancient communities of the world. The use of microorganisms for fermentation, domesticating animals for livestock, alcohol in the form of wine and beer, and herbal remedies and plant balms for the treatment of wounds and ailments are few examples of classical and ancient biotechnology. The term biotechnology was used before the twentieth century for traditional activities such as making dairy products, bread, or wine. Pre-twentieth-century renaissance accelerated the pace of scientific discoveries in Europe. Galileo, Copernicus, and Leonardo Da Vinci were some of the renaissance leaders whose knowledge extended over various branches of art, science, and medicine. Microscopes, first cork cell, protozoa, smallpox vaccine, and Darwin's theory[36] of evolution are some of the famous discoveries related to this very field during that time.

Modern Biotechnology

Modern biotechnology can be traced back to the eighteenth century when smallpox vaccine was used to invoke an immune response and prevent the development of more serious cases later in life. If you want a quick insight into what modern biotechnology is all about, start thinking of yourself as being composed of and run by molecules. Thanks to the cooperation of these small chemical units, we can blink, breathe, and read. It is because of molecules that we grow from microscopic fertilized eggs into functioning human beings. Biotechnology operates at

---

[35] Grubb, P.W. 1999. *Patents for Chemicals, Pharmaceuticals and Biotechnology: Fundamental of Global Law, Practice and Strategy*, pp. 224–5. Oxford, New York: Clarendon Press.

[36] Grace, Eric C. 1997. *Biotechnology Unzipped: Promises and Realities*, p. 7. Washington: Joseph Henry Press.

the molecular level where there is no real difference between a person and a bacterium. What is new about modern biotechnology is not the principle of using various organisms but the techniques for doing so. Darwin's theory of evolution makes two important points relevant to biotechnology. First, every species is ultimately related to one another through common ancestors, in spite of differences in physical manifestation. Second, the theory implies that a record of the evolutionary past is present inside every living thing. The field of genetics studies relationships between living beings. It also studies the intra- and inter-species relationships. Genetic study reveals the story and history of a living being including its evolution and genetic setup that it has received from its past generations. In early twentieth century, the real modern biotechnology movement started, particularly in immunology and genetics.[37] In fact, modern biotechnology is greatly influenced by two techniques that have been invented in the 1970s, namely, rDNA technology (genetic engineering) and hybridoma technology. The first technique is all about extracting genetic material from one living being and incorporating it into the intended other. It involves the insertion of genetic material into an intended cell from an external source for the production of a desired protein by the cell. The second technique is all about fusing different cells together to form a hybrid cell line producing certain types of antibodies useful in the production of medicines.[38]

Penicillin, computers, discovery of DNA as the genetic basis, and the use of bacteria to treat raw sewage (bioremediation project) have been some of the significant developments in the twentieth century.[39] A revolution in forensics and biomedical science took place with the new lab methods, such as DNA sequencing, protein analysis, and polymerase chain reaction. The millennium ended with the introduction of first cloned sheep Dolly, which led to a debate over the ethical issues relating to stem cell research, genetic testing, and genetically modified organisms. The twentieth century started with the development of a rough draft

[37] Sreenivasulu. 2005. 'Patenting Genetically Modified Life Forms'.
[38] Grubb. 1999. *Patents for Chemicals, Pharmaceuticals and Biotechnology,* p. 225.
[39] Sreenivasulu. 2007. 'New Face of Patent Law'.

of human genome, or map of human life. Nanotechnology, biotechnology, and cognitive and information sciences combine the knowledge of biotechnology generated in the last 30 years to develop abilities in computing advances (information technology), in manipulating matters at the atomic level (nanotechnology), and in understanding the human brain (cognitive sciences).

Origins and Development

Biotechnology in one form or another has flourished since prehistoric times. When the first human beings realized that they could plant their own crops and breed their own animals, they learned to use biotechnology. The discovery that fruit juices could be fermented into wine or that milk could be converted into cheese or yogurt, or that beer could be made by fermenting solutions of malt and hops began the study of biotechnology. In fermentation, microorganisms such as bacteria, yeasts, and molds are mixed with ingredients that provide them with food. As they digest this food, the organisms produce two critical by-products: carbon dioxide gas and alcohol. When the first bakers found that they could make soft, spongy bread rather than a firm, thin cracker, they were acting as fledgling biotechnologists. The first animal breeders, realizing that different physical traits could be either magnified or lost by mating appropriate pairs of animals, engaged in the manipulations of biotechnology.

The Humble Beginning

Biotechnology arose from the field of "zymotechnology", which deals with industrial fermentation. It began as a search for a better understanding of industrial fermentation, particularly beer which was an important industrial commodity and not only a social one. Certain practices that we would now classify as applications of biotechnology have been in use since the earliest days of humankind. Nearly 10,000 years ago, our ancestors were producing wine, beer, and bread using fermentation, a natural process in which the biological activity of one-celled organisms plays a critical role. The story of the use of biological systems for the fulfillment of human needs perhaps started in 6000 BC.[40] In fact, biotechnology can be traced back to approximately 6000 BC when the Sumerians and the Babylonians first used yeast to make beer.[41] In beer-making, yeast cells break down starch and sugar (present in cereal grains) to form alcohol; the froth, or head, of the beer results from the carbon dioxide that the cells produce. In simple terms, the living cells rearrange chemical elements to form new products that they need to live and reproduce. Coincidently, in the process of doing so, they help make a popular beverage. By 4000 BC, Egyptians discovered how to bake leavened bread using yeast. Other fermentation processes were established in the ancient world, notably in China. In China, fermentation processes were discovered for preserving milk to produce yogurt and cheese. The preservation of milk using lactic acid bacteria resulted in yogurt; molds were used to produce cheese; and vinegar and wine were manufactured by fermentation. Babylonians celebrated the pollination of date palm trees with religious rituals. The discovery of the fermentation process allowed people to produce foods by allowing living organisms to act on other ingredients. Our ancestors also found that, by manipulating the conditions under which fermentation took place, they could improve both the quality and the yield of the ingredients themselves. Bread-baking is also dependent on the action of yeast cells. The bread dough contains nutrients that these cells digest for their own sustenance. The digestion process generates alcohol (which contributes to wonderful aroma of baking bread) and carbon dioxide (which makes the dough rise and forms the honeycomb texture of the baked loaf). In cultivation, certain new developments were noticed by 1000 BC when Babylonians celebrated the pollination of date palm trees with religious rituals.[42] It seems by that time itself crossing-breeding of plants and pollination of trees were in practice. In 420 BC, Socrates[43] researched on heredity.[44] He speculated on why children do not always resemble their parents. He

[40] Ranga. 1999–2000. *Animal Biotechnology*, p. 1.

[41] Rockman, Havard B. 2004. *Intellectual Property Law for Engineers and Scientists*. New Jersey: I Wiley IEEE Press.

[42] See Sreenivasulu, N.S. 2007.'New Face of Patent Law'.

[43] Greek philosopher who lived between 470 and 399 BC.

[44] See Sreenivasulu. 2008. *Biotechnology and Patent Law*.

remarked that the sons of great statesmen were usually lazy and good for nothing. Furthermore, there was some research on heredity character- istics and traits that children receive from their parents. Hippocrates[45] determined that the male contribution to a child's heredity is carried in the semen. By analogy, he guessed there is a similar fluid in women, since children clearly receive traits from each in approximately equal proportion. The findings of Socrates and Hippocrates helped further scientific research on heredity. However, Aristotle[46] rejected the theories of Hippocrates. According to him, all inheritance came from the father. He asserted that the male semen determined the baby's form, while the mother merely provided the material from which the baby was made. He suggested that female babies were born due to "interference" from the mother's blood. Hindu philosophers first pondered the nature of repro- duction and inheritance between 100 AD to 300 AD. Research findings on inheritance and heredity continued, and in 1000 AD, Hindus observed that certain diseases may "run in the family." Moreover, they came to believe that children inherit all their parents' characteristics. According to the laws of Manu, "[a] man of base descents can never escape his origins." By 1400 AD, distillation of a variety of spirits from fermented grain was widespread. Meanwhile, religious flavor started to influence research. Few regions gave up research on nonnatural food and products as well as research on inheritance. However, few regions continued their research irrespective of religious objections on the inventions resulting from such research. Egypt and Persia largely gave up brewing as a result of the influence of Islam. However, fermented breads and cereals still maintained their hold in the African diet.

Early Developments

The seventeenth century witnessed certain important breakthroughs, but their credibility was questionable and was confirmed later by further researches.[47] In 1665 AD, Robert Hooke observed that living beings were

---

[45] Greek physician who lived between 460 and 377 BC.

[46] Greek philosopher who lived between 384 and 322 BC.

[47] For example, in 1630 AD, William Harvey discovered that plants and ani- mals alike reproduce in a sexual manner: males contribute pollen or sperm, and

made of units called cells.[48] However, it was not confirmed until almost 200 years later when scientists, armed with better microscopes, realized that all of us were made of very small compartments. In 1673 AD, Anton van Leeuwenhoek, a Dutch merchant and civic administrator, used a microscope to make discoveries in microbiology. He was the first scientist to describe protozoa and bacteria and to recognize the role played by microorganisms in fermentation. His work on the flea, wherein he proclaims that fleas—like fish, dogs, and humans—are sexual beings is considered to be a landmark. He also confirmed the discovery by Louis Dominicus Hamm of the existence of sperm cells. By 1675 AD, Marcello Malpighi discovered that the nervous system of living beings was a bundle of fibers connected to the brain by the spinal cord. He made detailed observations on the anatomy of the silkworm and described the development of the chick in its egg in his published work on plant anatomy. He made extensive use of the microscope to study blood circulation through the nervous system. In the eighteenth century, the empirical method and the industrial revolution brought monumental changes in farming and industry. During this time, the works by Darwin and Pasteur inspired research in biological sciences. At the same time, the microbial nature of many diseases was established and Mendel toiled in obscurity with his pea plants. In 1701, in Constantinople, Giacomo Pylarini practiced "inoculation," which involved intentionally giving children smallpox inoculum to prevent a serious case later in life.[49] A century later, inoculation became a very popular and extensively practiced method to prevent diseases. Almost after 100 years from the invention of inoculation, Edward Jenner published a book comparing vaccination (intentionally infecting humans with cowpox inoculum to induce resistance to smallpox) to inoculation (intentionally infecting humans with a putatively mild strain of smallpox to induce resistance to a severe strain of the disease). He derived his ideas from observing that people who had been exposed to cowpox were not

---

females contribute eggs. Louis Dominicus Hamm discovered that sperm exist in the form of cells. However, 200 years passed before the first mammalian eggs were observed.

[48]   See Grace. 1997. *Biotechnology Unzipped*, p. 3.

[49]   Inoculation competed with "vaccination"—an alternative method that uses cowpox rather than smallpox as the protecting treatment—for a century.

vulnerable to smallpox. The century spanning 1750 to 1850 witnessed research to enhance the yield in cultivation. The researchers also focused on the proper use of land in cultivation. In particular, the farmers in Europe increased their cultivation of leguminous crops and began rotating crops to increase yield and land use. Beginning with fermentation, the use of biological processes experienced many changes over centuries.[50] In 1859, Charles Darwin (1809–82), in his landmark book *On the Origin of Species*, hypothesized that animal populations adapt their forms over time to best exploit the environment, a process he referred to as "natural selection." As he traveled in the Galapagos Islands, he observed how the finch's beaks on each island were adapted to their food sources. He theorized that only the creatures best suited to their environment survive to reproduce. Darwin also inferred the process of adaptive radiation, wherein populations spread out into the environment to exploit specialized resources. In contrast to the ideas of Justis Liebig, Louis Pasteur asserted that microbes were responsible for fermentation. His experiments in the ensuing years proved that fermentation is the result of the activity of yeasts and bacteria. In the nineteenth century, Louis Pasteur, called the father of biotechnology, demonstrated the fermentative ability of microorganisms.[51] During that time, epidemiological observations were used to develop the hypothesis of cross-infection by spread of childbed fever, which led to the hypothesis that physicians should wash their hands after examining each patient. Another milestone during the nineteenth century was the invention of the process of beer fermentation by Louis Pasteur. In the ensuing years, he proved that the activity of yeasts and microbes were responsible for fermentation. In 1863, he invented the pasteurization process, which involves heating wine sufficiently to inactivate microbes that would otherwise turn the "vin" to "vin aigre" or "sour wine," while at the same time not ruining the flavor of the wine. In the 1940s, complicated techniques were introduced to the mass cultivation of microorganisms to exclude contaminated microorganisms. He theorized that decayed organisms were found as small organized "corpuscles" or "germs" in the air.

In 1865, Gregor Mendel[52] presented his laws of heredity to the Natural Science Society in Brunn, Austria. Mendel proposed that

[50] Ranga. 1999–2000. *Animal Biotechnology*, p. 1.
[51] Smith. 1996. *Biotechnology*, p. 4.
[52] Gregor Mendel (1822–84) was an Augustinian monk.

invisible internal units of information account for observable traits, and that these "factors," which later became known as genes, are passed from one generation to the next.[53] Mendel's work remained unnoticed, languishing in the shadow of Darwin's more sensational publication five years earlier, until 1900, when Hugo de Vries, Erich Von Tschermak, and Carl Correns published research corroborating Mendel's mechanism of heredity. In 1868, Fredrich Miescher, a Swiss biologist, successfully isolated nuclein, a compound that includes nucleic acid, from pus cells obtained from discarded bandages. Meischer, however, was not investigating heredity. Instead, he was trying to identify the chemicals in cells. Several generations of scientists would pass before the connection would be made between the DNA made up of nucleic acids[54] found by Miescher and the laws of heredity described by Mendel just three years earlier. In 1873, Joseph Lister described the "most probable number" technique, the first method for the isolation of pure cultures of bacteria, an important step in understanding infectious diseases. In 1883, August Weismann, a German physiologist coined the term "germplasm." He asserted in his book of the same name that both parents contribute equally to the heredity of the offspring, that sexual reproduction thus generates new combinations of hereditary factors, and that the chromosomes must be the bearers of heredity. His books were translated promptly into French and English. In 1884, Pasteur developed rabies vaccine, a much awaited invention to treat the disease. In the following years, human trials of the rabies vaccine were conducted. In 1887, Edouard-Joseph-Louis-Marie van Beneden discovered that each species had a fixed number of chromosomes; he also discovered the formation of haploid cells during cell division of sperm and ova (meiosis). In 1896, Wilhelm Kolle, a German bacteriologist, developed cholera and typhoid vaccines. In the following year, Eduard Buchner demonstrated that fermentation could occur with an extract of yeast in the absence of intact yeast cells. This was a founding moment in biochemistry and enzymology. In 1900, Walter Reed established that yellow fever is transmitted by mosquitoes. It was the first time a human disease was shown to be caused by a virus.

---

[53] See Grace. 1997. *Biotechnology Unzipped*, p. 8.
[54] See generally, Sreenivasulu. 2008. *Biotechnology and Patent Law*.

Middle Ages

In the late nineteenth century Germany, brewing contributed as much to the gross national product as steel, and taxes on alcohol proved to be a significant source of revenue for the government. In 1890s, institutes and remunerative consultancies were dedicated to the technology of brewing. The most famous was the private Carlsberg Institute, founded in 1875, which employed Emil Christian Hansen, who pioneered the pure yeast process for the reliable production of consistent beer. Lesser known were private consultancies that advised the brewing industry. One of these was the Zymotechnic Institute, which was established in Chicago by a German-born chemist John Ewald Siebel.[55] Necessity is the mother of invention, and necessity also rolls out innovative thoughts, perceptions, and approaches. During the First World War, when there was scarcity of food, raw material for industries in support of war influenced the growth and expansion of zymotechnology. The War had a lot to do with the advancement in zymotechnology. With food shortages spreading and resources fading, some dreamed of a new industrial solution. In Germany, Max Delbruk, through the fermentation process, grew yeast on an immense industrial scale during the war to meet Germany's animal feed needs. In the situation where there was scarcity of hydraulic fluid and glycerol, lactic acid a fermented product was used. In similar situations, in Russia, Chaim Weizmann used starch to eliminate Britain's shortage of acetone, a key raw material in explosives, by fermenting maize to acetone. The industrial potential of fermentation was outgrowing its traditional home in brewing and soon zymotechnology gave way to biotechnology.

Indeed the term "biotechnology" was not recently coined.[56] It was coined in 1919 by Karl Erkay, a Hungarian engineer.[57] At that time biotechnology was known as a technique used to produce desired goods from raw materials with the aid of living organisms. Karl Erkay described biotechnology as a technology based on converting raw materials into

[55] See Banerjee, Ritu and Gargi Mukherjee. 2002. 'Developments in Biotechnology: An Overview', *Indian Journal of Biotechnology*, 1(1): 9–16.

[56] Grubb. 1999. *Patents for Chemicals, Pharmaceuticals and Biotechnology*, p. 225.

[57] Rockman. 2004. *Intellectual Property Law for Engineers and Scientists*.

a more useful product. For him, biotechnology meant the process by which raw materials could be biologically upgraded into socially useful products. On the basis of his experiments in biotechnology research and development, he built a slaughterhouse for a thousand pigs and also a fattening farm with space for 50,000 pigs, raising over 100,000 pigs a year. He advocated that biotechnology could provide solutions to societal crises, such as food and energy shortages. Seemingly, from the experiences of the War, Erkay had foreseen that the future food shortages, energy shortages, and health concerns could be addressed through biotechnology. He strongly believed in the potential of biotechnology in catering the various needs of the society. Between 1900 and 1950, the two world wars killed millions and lead to large-scale destruction of property. However, during the same time efforts were made to explore the miracle and mechanism of reproduction: the nature and structure of DNA. When Mendel discovered that genetic characters were passed from parents to children and played a role in the designing of the features and characteristics of children, it was not well-received or completely accepted. Mendel felt that although his discoveries were not well appreciated, there would be a time when researchers would look back to his work. He firmly believed in his discoveries unleashing the secrets of human characteristics and the role played by genetic identity which passes from parents to children. Nevertheless, research continued on unleashing the secrets of human life. In 1900, Hugo DeVries, Erich Von Tschermak, and Carl Correns independently rediscovered Gregor Mendel's discoveries on inherited characteristics of children. This rediscovery formally signaled the birth of another sub-discipline in biotechnology named genetics. At the same time, William Sutton observed homologous pairs of chromosomes in grasshopper cells, which fuelled interesting debates on genetics, and further researches were encouraged and initiated. Post the Second World War, "biotechnology" not only entered German dictionaries but became a catchword. Hungarian business consultancies were instrumental in making the term popular not only across Europe but also in the United States (US).[58] The world war had given rise to circumstances that encouraged biological industries to create opportunities

---

[58] See also Rockman. 2004. *Intellectual Property Law for Engineers and Scientists.*

for new fermentation products. Fermentation technology and fermented products created more opportunities, growth, and development. There started a belief that the needs of the society could be met by fermentation technology. In particular, the needs of an industrial society could be met by fermenting agricultural waste. At the same time it was observed that major outbreaks of disease in overcrowded industrial cities led to the introduction of large-scale sewage purification systems based on microbial activity. It was also discovered that key industrial chemicals (glycerol, acetone, and butanol) could be generated using microorganisms such as bacteria. In 1902, Walter Stanborough Sutton stated that chromosomes were paired and might be the carriers of heredity. He suggested that Mendel's "factors" were located on chromosomes.[59] After observing chromosomal movements during meiosis, Sutton developed the chromosomal theory of heredity. Sutton noticed that chromosomes occur as pairs, and that gametes (egg and sperm cells) receive only one chromosome from each pair during meiosis. This corroborated Mendel's theory that the genetic "factors" were segregated. Sutton gave Mendel's "factors" the name we use today: "genes." At the same point of time in history, Archibald Garrod made the connection between Mendelian heredity and the biochemical pathways of reproduction in the individual organism. He went on to inspire a distinguished line of researchers whose work in the ensuing 90 years became indispensable to the growth of human genetics. In 1903, Walter Sutton and Theodor Boveri, working independently, proposed that each egg or sperm cell contains only one of each chromosome pair.[60] In 1905, it was discovered by Edmund Wilson and Nellie Stevens that two different types of chromosomes, namely, "X" and "Y," determine sex. It was observed that the Y chromosome represents the male sex, and the X chromosome represents the female sex. In determining the sex

[59] According to Mendel, human characteristics flow from some factors in the human cell that are located on the chromosomes. According to recent research, chromosomes are made up of DNA and the functional part of DNA is known as gene. Mendel probably pointed to genes when he said that there are factors flowing from chromosomes that are responsible for human characteristics.

[60] This connected two phenomena: the patterns by which pairs of Mendel's factors assort themselves and the precisely similar sorting and recombination of the chromosomes in the formation of the germ cells and the fertilization of the egg.

of a fetus, a single Y chromosome and a single X chromosome together determine maleness, and two copies of the X chromosome determine femaleness. However, it seems that all the characteristics are not inherited from parents; there can be certain characteristics that are unique to the children or the respective generation. Each generation may acquire certain special or independent features, which may or may not pass to the next generation. It cannot be believed that children inherit all the characteristics from their parents. They may possess certain traits and abilities that parents do not have. At the same time, whatever characteristics parents possess need not be reflected in children. On these lines, Thomas Hunt Morgan, in 1911, explained the separation of certain inherited characteristics that are usually linked to be caused by the breaking of chromosomes sometime during the process of cell division. Morgan began to map the positions of genes on the chromosomes of the fruit fly. Therefore, transfer of characteristics depends on the division of cells carrying the characteristics. Since the characteristics of both mother and father come at crossroads at the time of fertilization, it could be asserted that at the time of fertilization when the egg and the sperm meet, characteristics get crystallized into the fetus. Going further in genetic research, William Bateson and Reginald Crudell Punnett, between 1905 and 1908, demonstrated that some genes modify the action of other genes. These findings are used nowadays to suppress the functioning of any gene in the genome or in modifying the functioning of desired genes. In fact it was not until 1909, when the term gene was coined, that the factors responsible for carrying and transferring genetic characteristics were known as factors of characteristic features. In 1909, Wilhelm Johannsen coined the terms "gene" to describe the carrier of heredity; "genotype" to describe the genetic constitution of an organism; and "phenotype" to describe the actual organism, which results from a combination of the genotype and the various environmental factors. In the same year, Phoebus Levene discovered that the sugar ribose is found in some nucleic acids, which we now call RNA. The following year witnessed a major breakthrough regarding the location of genes.

Expansion of Boundaries

Thomas Hunt Morgan proved that genes are carried on chromosomes, establishing the basis of modern genetics. With his co-workers, he demonstrated the location of various fruit fly genes on chromosomes, establishing the use of Drosophila fruit flies to study heredity. Morgan's group also demonstrated the existence of sex-linked genes, and over the next 10 years, expanded the idea to other trait linkages, using "crossing-over" to help determine the location of genes. A big leap in the genetic research took place when Herbert M. Evans, in 1918, discovered that a human cell contains 48 chromosomes. In 1920, a Bureau of Biotechnology was established in Leeds, UK, which developed expertise in the fermentation process. The Bureau published a journal dealing with fermentation technology and related topics.[61] At this juncture, research on fermentation technology was developing; in particular, fermented nonalcoholic drinks started to hit the market. The Bureau of Biotechnology offered expertise in fermented nonalcoholic drinks.[62] Fermentation-based processes and products generated out of such processes showed enormous and ever-growing utility, which had a great impact on the society. The growth and influence of fermentation technology spread across Europe and resulted in the invention of penicillin in England. In fact, Gregor Mendel's presentation of the law of heredity allowed humankind to know about heredity and genetic follow-up from one generation to the next generation in detail.[63] He discovered the law of heredity, the statistical relationships that govern how characteristics are passed from one generation to the next. One of the most important results of Mendel's work was his demonstration that inherited characteristics are determined by genes that pass over generations. He also inferred that each organism contains two copies of each factor, one inherited from its mother and the other from its father.[64] Perhaps the development of the rabies vaccine was an addition to the armory of medical sciences to find cures to different diseases through biotechnology and biomedical research. Furthermore, the invention of penicillin by Alexander Fleming during the times of world war played a major role in the advancement of biotechnology. Penicillin, being the first antibiotic,

[61] Grubb. 1999. *Patents for Chemicals, Pharmaceuticals and Biotechnology*, p. 225.

[62] See Banerjee and Mukherjee. 2002. 'Developments in Biotechnology: An Overview'.

[63] Smith. 1996. *Biotechnology*, p. 4.

[64] Grace. 1998. *Biotechnology Unzipped*, p. 8.

revolutionalized the pharmaceutical industry. It quickly spread across the US, and industrial production of penicillin started in Illinois. The potential benefits of penicillin and the health needs that it serves brought radical changes in the society's perception toward it. Penicillin became a hope for the needy and it was considered the miracle drug. Public expectations from penicillin increased, and because of its increased use, there were enormous profits, which in turn influenced the growth of biotechnology and its research. Between 1920 and 1930, techniques of biotechnology started gaining popularity due to their potential. Especially in the US, the growth and use of new techniques to improve the yield of agriculture were on upsurge. Hybridization and plant breeding became popularized in the US, greatly improving the productivity of agriculture.[65] Meanwhile, in 1921, Hermann J. Muller observed the nature of the gene to be astonishingly prescient. He conceived of the gene as a particle that, despite its ultramicroscopic size, exhibits a complex structure of different parts. At the time when biotechnology was showing a lot of promise in catering the different needs of the community, the field of politics was not far away from the influence of biotechnology. In fact, biotechnology and genetics were also used in politics and administration for different reasons. The eugenics movement influenced the US congress to pass the US Immigration Act of 1924, limiting the influx of poorly educated immigrants from Southern and Eastern Europe on the grounds of suspected genetic inferiority. In 1926, Hermann Muller discovered that x-rays induce genetic mutations more quickly than under normal circumstances. This discovery provided researchers with a way to induce mutations, an important tool for discovering what genes do on their own. Muller made his findings based on his study on fruit flies. His study revealed that x-rays induce genetic mutations in fruit flies 1,500 times more compared to normal mutations. Laying foundations to genetic engineering in future to alter and modify the structure and performance of a living organism by modifying the structure of genes, in 1928, Fredrick Griffiths noticed that a rough type of bacterium changed to a smooth type when an unknown "transforming principle" from the smooth type was present. Sixteen years later, Oswald

---

[65] Plant breeding efforts started in 1943 in search of better crops with maximum yield making Mexico self-sufficient in wheat production for the first time.

Avery identified that "transforming principle" as DNA.[66] The discovery of DNA started a completely different field of science called genetics. Genetics is a study related to the genetic makeup of a living being and functioning of such genetic makeup.[67] It deals with the impact on the structure of a living being based on its genetic makeup that passes from one generation to another. Research in genetics empowered the human-kind to manipulate the genetic makeup known as genetic engineering of the living being, which is otherwise received by the living being from its past generation. By 1928, the age of penicillin began, followed by the discovery of antibiotics in the subsequent year. Although it was almost 15 years before they were made available to the community for medicinal use, the discovery of antibiotics and penicillin were major breakthroughs in the history of biotechnology. The production of antibiotics is based on the isolation of products from selected strains of microorganisms through fermentation.[68] Although a majority of antibiotics are now produced synthetically, many are still made from microorganisms either found in nature or artificially mutated. The potential and significance of antibiotics made them popular in no time and their use in medical field on a regular basis mandated their industrial production. Eventually, in the 1940s large-scale production of antibiotics began, and by 1945, large-scale production of penicillin was achieved. In 1929, Phoebus Levene discovered a previously unknown sugar, deoxyribose, in nucleic acids in the chromosomes that do not contain ribose; those nucleic acids are now known as deoxyribonucleic acids, or DNA. Current research reveals that DNA is the building block of living organisms; altogether the features of an organism are coded and decided at the DNA level. In 1935, Andrei Nikolaevitch Belozersky isolated DNA in the pure state for the first time which led to researches involving the incorporation of foreign DNA into an organism. Since researches at the micro level were on the upsurge with the discovery of DNA, the functioning, isolation, and purification of DNA were all conducted at the micro level; there

---

[66] For more details, see Rockman. 2004. *Intellectual Property Law for Engineers and Scientists.*

[67] See generally, Sreenivasulu, N.S. 2008. *Biotechnology and Patent Law.*

[68] Grubb. 1999. *Patents for Chemicals, Pharmaceuticals and Biotechnology,* p. 225.

emerged a new discipline in life sciences called microbiology. In 1938, the term "molecular biology" was coined. In 1944, Oswald Theodore Avery, Colin MacLeod, and Maclyn McCarty determined that DNA is the hereditary material responsible for carrying the hereditary characteristics from one generation to another. However, this theory gained little attention to begin with because scientists believed that DNA was too simple a molecule to the complete genetic information for an organism. In the same year, Barbara McClintock, working in Cold Spring Harbor, New York, discovered that genes can be transposed from one position to another on a chromosome.[69] This discovery has a real impact on the microbial and genetic research, since it revealed the possibility of modifying the genetic structure of chromosomes as desired. The discovery of McClintock was rightly honored with Nobel Prize as it vested the power of changing the natural structure of a living organism in the hands of human beings. Her invention virtually initiated human dominance over nature. By 1945, the isolation of animal cell became a reality and isolated animal cell cultures in laboratories began. In the same year, it was discovered how genetic information is transferred. Furthermore, during the same time more complex techniques were introduced to the mass cultivation of microorganisms to exclude contaminated microorganisms.[70] In 1946, Edward Tatum and Joshua Lederberg showed that bacteria sometimes exchange genetic material directly by a process that they called conjugation. In fact, living organisms do exchange genetic material naturally over a period of time. The theory of natural selection also states that living organisms do leave certain characteristics naturally over a period of time and vice versa. In the same year, Max Delbruck and Alfred Day Hershey independently discovered that the genetic material from different viruses can be combined to form a new type of virus. This process was another example of genetic recombination. In 1947, Barbara McClintock first reported on "transposable elements," known today as "jumping genes." The scientific community failed to appreciate the significance of her discovery at the time. Genes naturally jump from one organism to other; in fact genes incorporated through genetic

---

[69] McClintock was awarded the Nobel Prize in Physiology or Medicine in 1983 for the discovery.
[70] Smith. 1996. *Biotechnology*, p. 4.

engineering can also jump from the host organism to any other. In 1950, it was discovered by Erwin Chargaff that in DNA, adenine, thymine, guanine, and cytosine are present in equal capacities. These relationships were later known as "Chargaff's rules" and served as a key principle for Watson and Crick in assessing various models for the structure of DNA. In the same year, artificial insemination of livestock using frozen semen was conducted for the first time. Artificial insemination of livestock could increase the yield for the farmers. In the 1950s, steroids were invented. Steroids were synthesized using fermentation technology and were promising in fulfilling the health needs of the society. They also revolutionized the medical sciences as penicillin. In particular, the steroid cortisone had great potential and became a milestone in the history of biotechnology at large. In 1952, Joshua Lederberg and Norton Zinder showed that bacteria sometimes exchange genes by an indirect method, which they termed "transduction," in which a virus mediates the exchange by taking bits of DNA from one bacterial cell and transporting the bacterial genes into the next cell it infects. In the same year, William Hayes discovered conjugation, the process whereby one bacterial cell transports a copy of some of its genes into a second bacterial cell. It is a natural process by which a bacterium transfers some of its genes from one cell to another. This discovery established the fact that transferred genes could survive in the new atmosphere and function according to the directions of the host cell. Besides, the invention of electron microscopy showed the inside of cells to be filled with minute but well-formed anatomical structures, including a vast number of a complex molecular organ now termed as the ribosome. With this, scientists were able to see though the complicated cell structure and its physical and chemical properties. The discovery of the structure of DNA and its isolation by Crick and Watts in 1953[71] started a new era in biotechnology.[72] Rosalind Franklin contributed a lot in the discovery of the structure of DNA, but did not receive the expected limelight. Crick and Watson proposed the double-stranded, helical, complementary, anti-parallel model for DNA. Isolation and purification of DNA led to the possibilities of incorporation of purified DNA into any intended organisms for desired results. Similarly,

---

[71] Grace. 1998. *Biotechnology Unzipped*, p. 15.
[72] See Sreenivasulu, N.S. 2008. *Biotechnology and Patent Law*.

isolation, purification, and incorporation into an intended cell, of the functional portions of DNA,[73] which we call gene today, also became possible after the discovery by Crick and Watson. Such incorporation of isolated and purified DNA/gene from one organism to another requires some carriers that will carry the DNA/gene and place the same in the intended location of the chromosome of the host organism.

Genetic Research: Challenging the Nature

The path of genetic manipulation can be said to have started in 1665 when the English scientist Robert Hook discovered that living beings are made up of a tiny biological material called "cell." Ten years later Anton van Leeuwenhoek became the first person to observe and describe microorganisms. Later many scientists could find cells in every part of both plants and animals. However, in 1839 two German biologists, Matthias Schleiden and Theodore Schwann, propounded "cell theory," which says that all organisms are made of cells.[74] Some organisms such as bacteria have only a single cell, while others such as plants, animals, and human beings have multiple cells in their biological setup. Organisms having only a single cell in their biological setup are known as unicellular organisms, and those with more than one cell in their biological setup are known as multicellular organisms. The life of even a multicellular organism starts from a single cell and grows through the multiplication of the single cell into a complete plant, animal, or human being. A cell contains all information needed to build the organism. Such information is passed from one generation to another. It was later discovered that a cell contains a nucleus, which is the driving force of the cell, containing two sets of chromosomes. These chromosomes are in the shape of double helix containing the information regarding the makeup of the cell and the organism containing the cell. It was discovered that every species has a specific number of chromosomes inherited from the parents. For example, a human cell contains 46 chromosomes, 23 each coming from the mother and the father. If an organism has got more

---

[73] See also Rockman, H.B. 2004. *Intellectual Property Law for Engineers and Scientists*, p. 259. NJ, USA: Wiley–Interscience.

[74] Grace. 1998. *Biotechnology Unzipped*, pp. 3–6.

than one cell, every cell will have the same number of chromosomes inside the nucleus, except the reproductive cells. Reproductive cells contain half the number of chromosomes that a regular cell contains. When male and female reproductive cells meet and fertilize, the new cell that forms will contain 46 chromosomes. The fertilized cell gets heredity information equally from the mother's and father's reproductive cells. In 1869, Johann Miescher, a Swiss scientist, tried to explore the content and the interior of a cell. He discovered that there is some chemical inside the nucleus, which he named as the nucleic acid. He also discovered "deoxyribonucleic acid (DNA)." The functional part of DNA is known as gene,[75] which has no separate existence. In fact gene is nothing but a fragment of DNA with a specific function.

Noble Prize Boosts Research in Biotechnology

A famous science journal *Nature* published the manuscript of James Watson and Francis Crick describing the double helix structure of DNA in 1953. In the same year, William Hayes discovered that in the process of genetic manipulation, plasmids can be used to transfer genetic markers from one bacterium to another. In 1957, Francis Crick and George Gamov worked out the "central dogma," explaining how DNA functions to make proteins. Their "sequence hypothesis" posited that the DNA sequence specifies the amino acid sequence in a protein. They also suggested that genetic information flows only in one direction, from DNA to messenger RNA to protein, the central concept of the central dogma. DNA has two major functions in the life of a cell: expression and replication. Expression is the general process by which proteins coded for by particular DNA sequences called structural genes are constructed when needed. Replication is the process by which copies of the DNA sequence are transferred to daughter cells.[76] In the same year, Matthew Meselson and Frank Stahl demonstrated the replication mechanism of DNA. In 1959, Francois Jacob and Jacques Monod established the

---

[75] See Sreenivasulu, N.S. 2005. 'Patenting Genetically Modified Life Forms: Legal Issues and Challenges', *Indian Bar Review*, XXXII(3&4).

[76] Cooper I.P. 1999. *Biotechnology and the Law*, pp. 1–11. New York: West Group.

existence of genetic regulation—mappable control functions located on
the chromosome in the DNA sequence, which they named the repres-
sor and operon. They also demonstrated the existence of proteins that
have dual specificities. During the 1960s, biotechnology seemed to be
the answer for the hunger and starvation in the world[77] when the process
of raising single cell protein was invented. This process opened the possibilities of growing microorganisms on oil that captured the imagination
of scientists, industrialists, and policymakers. During the times when the
entire world was facing the crises of hunger and starvation, the process
of raising single cell protein promised production of food at a low cost by
using waste. Eventually scientists and industrialists started investing their
time, energy, and money in the industrialization of the process. Major
companies such as British Petroleum invested huge amounts of money
by building a pilot plant in Sothern France at Cap de Lavera. Raising
proteins from a single cell was another achievement for biotechnology.
However, the process of single cell protein received a mixed response,
as there were cultural objections to it. The great contribution to the
development of biotechnology by way of the discovery of the structure
of DNA[78] was firmly honored with Noble Prize in 1962. Watson and
Crick shared the Nobel Prize for Physiology and Medicine with Maurice
Wilkins. Unfortunately, Rosalind Franklin, whose work greatly contributed to the discovery of the double helical structure of DNA, died before
this date, and the Nobel Prize rules did not allow a prize to be awarded
posthumously. In 1965, Harris and Watkins successfully fused mouse
and human cells, another milestone in the development of biotechnology. Their work ignited researches on the fusion of different cells of
humans and animals, humans and plants, and plants and animals. The
field of cell culture and tissue culture also received enough boost with
the achievement of Harris and Watkins. Going further, in 1967, Mary
Weiss and Howard Green took a crucial step in human gene mapping
with the publication of a technique for using human cells and mouse
cells grown together in one culture. This was called somatic cell hybridization. Further, in the 1970s, when energy crisis followed food crisis,

[77] See Sreenivasulu. 2006. 'Information Technology and Biotechnology'.
[78] See also Rockman. 2004. *Intellectual Property Law for Engineers and
Scientists*.

biotechnology seemed to provide the remedy; however, its potential was not properly recognized, realized, or utilized. Nevertheless, the greatest revolution of biotechnology took place in the 1970s and the 1980s when the product of interaction between the science of biology and technology came into wider existence and the relationship got the name biotechnology.[79] What may be described as modern biotechnology as distinct from the classical fermentation technology began in the 1970s with the two basic techniques of rDNA technology and hybridoma technology. In the first of these, also referred to as gene splicing or genetic engineering, genetic material from an external source is inserted into a cell in such a way that it causes the production of a desired protein by the cell; in the second, different types of immune cells are fused together to form a hybrid cell line producing monoclonal antibodies.[80] In 1972, the first ever rDNA was invented. Paul Berg isolated and employed a restriction enzyme to cut DNA. Berg used ligase to join two DNA strands together to form a hybrid circular molecule. The DNA was combined and composed using strands of different DNAs. This was the first rDNA molecule. After the discovery of the first-ever rDNA in 1962, in the same year in California the first-ever successful DNA cloning experiments were performed. By 1972, the means to splice the DNAs from different genomes was achieved by two groups in the biochemistry department of Stanford University.[81] In a letter to *Science*, a famous journal in the field of science, a Stanford biochemist Paul Berg and others called for the National Institutes of Health to enact guidelines for DNA splicing. Their letter recommended that scientists stop performing certain types of rDNA experiments until the questions of safety could be addressed. This letter was provoked by experiments planned by Berg, which had drawn vocal concern from the scientific community. Their concerns eventually led to the 1975 Asilomar Conference, where the safety in DNA experiments was a issue. This instance exhibited the growing frequency of DNA research and its potential in catering different needs. Scientists

---

[79] Ranga. 1999–2000. *Animal Biotechnology*, p. 1.

[80] Grubb. 1999. *Patents for Chemicals, Pharmaceuticals and Biotechnology*, p. 225.

[81] Kornberg, Arthur. 1995. *The Golden Helix, Inside Biotech Venture*, p. 28. Sausalito, CA: University Science Books.

anticipated experiments in DNA splicing, rDNA, and DNA cloning to lead to the transfer of DNA from one life form to another. Genetic engineering became a reality when a manmade gene was used to manufacture a human protein in bacteria for the first time. Biotechnology companies and universities were off to the races, and the world would never be the same again. In 1978, in the laboratory of Herbert Boyer at the University of California at San Francisco, a synthetic version of the human insulin gene was constructed and inserted into the bacterium *Escherichia coli*. Since that key moment, the trickle of biotechnological development has become a torrent of diagnostic and therapeutic tools, accompanied by ever faster and more powerful DNA sequencing and cloning techniques. Stanley Cohen, Annie Chang, and Herbert Boyer "spliced" sections of viral DNA and bacterial DNA, creating a plasmid with dual antibiotic resistance. They then spliced this rDNA molecule into the DNA of a bacterium, thereby producing the first rDNA organism. Cohen and Boyer's invention of the technique of engineering DNA has taken biotechnology to its greatest heights. Cohen and Boyer successfully cut a section of DNA; they cut a gene from a bacterium and incorporated into the DNA of another bacterium. They were successful in activating the incorporated gene in the engineered bacterium.[82] This enabled the bacterium to adopt and perform the function of the incorporated gene. Genes are nothing but functional portions of DNA. The function of a cell is determined by its genome and the chromosomes inside it. Since chromosomes code for DNA, their function depends on the structure and sequence of DNA. Perhaps the entire portion of DNA may not have functions influencing the overall functioning of the cell. According to the scientific reports, the function of 97 percent of the DNA is yet to be discovered. It is only three percent of DNA that does perform some function and is called as gene.

Manifestation of Life

Biotechnology is termed as the science that manifests life. Mythologically speaking, our fate is in the stars, but biotechnologically speaking, our fate

---

[82] Vemuganti, G.K. and D. Balasubramanian. 2002. 'Heralding the Dawn of Cultured Adult Stem Cell Transplantation', *Indian Journal of Biotechnology*, I(January): 40.

is in our genes. It is believed that there is a corresponding gene for every feature, characteristics, and quality a living being possesses. Biotechnology, being capable of modifying and marshalling the genes in living beings, has the potential to change the features, characteristics, and qualities of living beings. With such potential, biotechnology can alter our life and change our approach toward life. Therefore, it might not be untrue to say that biotechnology manifests life. Perhaps, the commercial production of human insulin through rDNA technology in the US showcased the potential of rDNA technology.[83] To alter the functioning of any living organism it is ultimately the genes inside the cells of that organism which would need to be engineered or altered. That is why the process of modifying the function of an organism is done at the gene level and is known as genetic engineering. The discovery of the function of DNA, the discovery of gene as the functional part of DNA, and invention of techniques for engineering the genes to isolate and transfer them from one organism to another have yielded incredible benefits.[84] There are several functions that the genes perform in the body of living organisms. These functions range from coordination and supervising the day-to-day activities of a living organism to combating and fighting against diseases. There are corresponding genes that are responsible for every activity that an organism performs. For instance, if an organism has to digest what it has eaten, there would be a gene that is responsible for releasing required solutions and chemicals for the digestion of food. If an organism is suffering from some disease, the genes that are responsible for fighting against diseases would naturally get activated and release some proteins and medicinal solutions into the body. Having said that, if we know that a particular gene is responsible for releasing proteins and solutions having medicinal values to fight against a disease, through genetic engineering it is possible to isolate that particular gene

[83] Research on recombinant human insulin was initiated in the early 1980s and resulted in the commercial production of human insulin by the incorporation of the gene coding for human insulin into bacteria. Since bacteria replicates rapidly, the insertion of gene into bacteria and its expression could yield a high production of insulin.

[84] See also Rockman. 2004. *Intellectual Property Law for Engineers and Scientists*.

from the body of the living organism and incorporate the same into an intended organism either to fight against the disease or for producing such proteins and solutions having medicinal values at larger scales. In fact a bacterium is known for quick replication and multiplication into many. The cells inside a bacterium replicate and divide quickly. Since genetic engineering is done at the gene level inside a cell, if a gene coding for a particular solution or protein having medicinal value is incorporated into the cell of a bacterium, through replication it multiplies manifold. In this way, some proteins and solutions that genes produce inside any organism having medicinal value can be produced on a commercial basis through genetic engineering, which is also known as rDNA technology.[85] It is called so because it recombines or restructures the sequence of an intended DNA. Genetic engineering became a reality when a manmade gene was first used to manufacture a human protein in a bacterium. In 1974, scientists were successful in expressing a foreign gene in a bacterium inserted through rDNA/genetic engineering method. Stanley Cohen and Herbert Boyer demonstrated the expression of a foreign gene implanted in bacteria by rDNA methods. Cohen and Boyer showed that DNA can be cut and reproduced by inserting the rDNA into an intended bacterium.[86] These researches showed a lot of promise and wealth leading to insertion of foreign rDNA into plants and animals to get desired results. In 1975, Kohler and Milstein fused cells together to produce monoclonal antibodies. Following enormous research in the field of rDNA and genetic engineering in 1976, the National Institute of Health in the US released the first guidelines for rDNA experimentation. The guidelines restricted many categories of experiments where safety was an issue. In 1977, Genentech, Inc., a pharmaceutical company in the US, reported the production of the first human protein manufactured in bacteria: somatostatin, a human growth hormone-releasing inhibitory factor. For the first time, a synthetic, recombinant gene was used to clone a protein. Many consider this to be the advent of the age of biotechnology. The introduction of as many as 16 bills in the congress to regulate rDNA[87] research reflected

---

[85] See Sreenivasulu. 2007. 'New Face of Patent Law'.

[86] See also Sreenivasulu. 2008. *Biotechnology and Patent Law*.

[87] See also Rockman. 2004. *Intellectual Property Law for Engineers and Scientists*.

the increase in research in this field. However, it also raised certain issues regarding the safety of such research as well as ethical issues involved in it. Addressing the safety of DNA research, the bills called for the development of bacteria and plasmids that could be prevented from escaping from the laboratory environment. However, none of the bills were passed. Nevertheless, it marked the congress's concern over the regulation of biotechnology and safety in DNA research. Meanwhile, the scientists were busy in finding new ways and means of sequencing DNA and conducting experiments of genetic engineering in different living beings. Subsequently, in 1978, Genentech, Inc., and The City of Hope National Medical Center announced the successful laboratory production of human insulin using rDNA technology. Genetic engineering touched new heights when Stanford University[88] scientists successfully transplanted a mammalian gene. At the same time, researchers from the Harvard University used genetic engineering techniques to produce rat insulin. Researches were successful in introducing a human gene into a bacterium for commercial production of the proteins that the gene codes for. During the 1960s and the 1970s, research in biotechnology was at its peak in the US. Industries, academia, as well as the state were contributing to thorough research and development in the field of biotechnology with mutual understanding and cooperation. Compared to other regions in the US, support from the state was very much felt in biotechnology research and development. The approach of the state in the US was to encourage anything that had scientific, technical, and commercial potential and viability. The US congress, executive in the form of patent office, and judiciary were very much encouraging research and development in biotechnology. Perhaps, modern biotechnology has got its roots in the early 1980s when rDNA technology or genetic engineering developed.[89] Modern biotechnology is mostly concerned with and concentrated on genetic engineering. In fact, the origins of biotechnology culminated with the birth of genetic engineering. In 1980, the Supreme Court of America gave a judgment, whereby it directed the US

---

[88] See for greater details, Kornberg. 1995. *The Golden Helix, Inside Biotech Venture*, p. 29.

[89] See generally, Rockman. 2004. *Intellectual Property Law for Engineers and Scientists*.

Patent Office to grant patents on "anything under the sun made by man." Deciding a case involving patenting of genetically modified microorganisms, it held that scientific and technological revolutions should be promoted and encouraged. Given the potential of biotechnology and its research in catering the needs of the community at different levels, it viewed that genetically engineered microorganisms comprising intellectual labor of the scientists deserved patent protection. The court considered the potential of microorganisms capable of cleaning up oil spills in combating environmental pollution. It was satisfied with the capacity of microorganisms installed through genetic engineering to earn a patent. This decision was a great achievement in the history of biotechnology, since after the decision, the US Patent Office granted at least 500 patents on genetically engineered microorganisms and other inventions of genetic engineering. Furthermore, this landmark judgment encouraged the biotechnology industry to put further efforts in research and development. Eventually Kary Mullis and others at Cetus Corporation in Berkeley, California, invented a technique for multiplying DNA sequences in vitro by polymerase chain reaction (PCR). PCR was called the most revolutionary technique in molecular biology in the 1980s. Cetus patented the process, and in the summer of 1991, sold the patent to Hoffman-La Roche, Inc., for USD 300 million. Researches in isolating DNA[90] and transferring genes from one organism to another yielded great results when a genetically engineered animal was produced. In 1981, scientists at Ohio University produced the first transgenic animal by transferring genes from other animals into mice. Historically, this mouse was the first-ever genetically engineered, nonnatural, and man-made animal.[91] In the process of commercial production of human insulin, the corresponding gene coding for insulin inside the human body was identified, isolated, and incorporated into a bacterium that, through replication and multiplication, produced it on a commercial scale. In 1982, genetically engineered human insulin got market approval for human use. Genentech, Inc. received approval from the Food and

---

[90] See generally, Rockman. 2004. *Intellectual Property Law for Engineers and Scientists.*

[91] By 1983, genetically engineered plants and their patenting in the US became a reality.

Drug Administration to market genetically engineered human insulin.[92] More recently, the techniques of genetic engineering have been applied to higher organisms to produce transgenic animals (through incorporation of genes of one animal into another animal) and plants (through incorporation of genes of one plant into another) and even to humans for the purpose of gene therapy. Gene therapy is a method of combating genetic or hereditary diseases. As mentioned earlier, for every activity and feature of a living organism, there lies a corresponding gene. If a person is suffering from a hereditary disease, genes exist corresponding to that disease that actually express some solutions in the body making the body suffer from the disease. This led to the invention of a process/method to identify and isolate genes that cause such diseases to make the person free from such genetic or hereditary diseases; the process is known as gene therapy.[93] Through this process, the genes would be either removed or suppressed from functioning. The scientists developed gene therapy as a method to treat hereditary diseases by 1985. The National Institute of Health in the US issued guidelines for performing experiments in gene therapy on humans.[94] Through genetic engineering it is possible to replace missing or defective genes coding for a protein required by the body or introduce genes into cancer cells that will render them easier to kill. We have learned a great deal about the different organisms that our ancestors used so effectively. The marked increase in our understanding of these organisms and their cell products give us the ability to control the many functions of various cells and organisms. Using the techniques of gene splicing and rDNA technology, we can now actually combine the genetic elements of two or more living cells. Functioning lengths of DNA can be taken from one organism and placed into the cells of another organism. As a result, the bacterial cells can be made to produce human molecules. Cows can produce more milk with the same amount of feed. Moreover, we can synthesize therapeutic molecules that have never existed before. In the contemporary world, in criminal justice administration, DNA fingerprinting and Norco analysis

---

[92] In 1982, the US Food and Drug Administration approved the first genetically engineered drug, a form of human insulin produced by bacteria.

[93] Sreenivasulu. 2007. 'New Face of Patent Law'.

[94] See also Sreenivasulu. 2008. *Biotechnology and Patent Law*.

assist the prosecution in identifying the culprits. Prosecution is blessed with this technology since 1984 when Alec Jeffreys introduced the technique of DNA fingerprinting to identify individuals; its successful practice became a reality in 1985 when DNA fingerprinting was used in court rooms in justice administration. In the same year, the scientists could successfully incorporate genetically engineered microorganisms into plants, which could promise us genetic modification of plants for better results.

Capable of Catering Different Needs of the Community

In 1980s, genetically engineered plants resistant to insects, viruses, and bacteria were field-tested for the first time. In 1986, the release of the first genetically engineered crop, the gene-altered tobacco plants was approved. Genetic engineering and transfer of genes could yield miracles when scientists developed a method to treat genetic diseases by shuffling genes. During 1986–7, the application of biotechnology brought developments in the field of medicine. Research in producing biopharmaceuticals was at an upsurge, and Chiron Corporation was granted licence on the first ever recombinant/genetically engineered vaccine for hepatitis. In 1987, Genentech received approval for using the recombinant method of treating heart attacks. Methods for treating various diseases including kidney failure and several other diseases were invented and tested during this time. In 1988, Harvard molecular geneticists Philip Leder and Timothy Stewart were awarded the first patent for a genetically altered animal, a mouse highly susceptible to breast cancer. In the same year, SyStemix, Inc. received a patent for the SCID-hu mouse, an immune-deficient mouse with a reconstituted human immune system. The mouse was engineered for AIDS research. Further, in 1995, a new transgenic mouse carrying a gene for human Alzheimer's disease was developed. In 1990, the first successful field trial of genetically engineered cotton plants was conducted by Calgene, Inc. The plants had been engineered to withstand the use of the herbicide bromoxynil. In the same year, Mary-Claire King, epidemiologist at University of California, Berkeley, reported the discovery of the gene linked with breast cancer in families with a high degree of incidence of the disease before the age of 45 years. Subsequently analyzing chromosomes from women in cancer-prone families, Mary-Claire

King, found evidence that a gene on chromosome 17 causes the inherited form of breast cancer and also increases the risk of ovarian cancer. Similarly, the gene associated with Parkinson's disease was discovered in 1996, which provided an important new avenue of research into the cause and potential treatment of the debilitating neurological ailment. Further in the same year, GenPharm International, Inc. created the first transgenic dairy cow. The cow was used to produce human milk proteins for infant formula. After five years from the National Institute of Health issuing guidelines on testing gene therapy on human beings, the first gene therapy took place on a four-year-old girl with an immune-system disorder called adenosine deaminase (ADA) deficiency. The therapy appeared to work, but set off a fury of discussion on ethics both in academia and in the media. Besides, the Human Genome Project, the international effort to map all of the genes in the human body, was launched with an estimated cost of USD 13 billion. The project was concluded in 2002 by mapping the entire genome of human beings and the same is being used for discovering the causes of different genetic disorders, diseases, and their treatments. In 1991, research on cancer with a goal to identify the causes for the disease yielded good results. In particular, the scientists could identify the gene responsible for breast cancer in females by analyzing the chromosomes in the cancer-prone family. In the following year, causes for ovarian cancer were discovered. By 1994, a number of genes responsible for various diseases and ailments were discovered.[95] The discovery of genetic material such as genes and DNA gave raise to certain proprietary issues. The year also saw an increase in squabbling

---

[95] A multitude of genes—human and otherwise—were identified and their functions described. These included:

- *Ob*: a gene predisposing to obesity.
- *BCR*: a breast cancer susceptibility gene.
- *BCL-2*: a gene associated with apoptosis (programmed cell death).
- hedgehog genes (so named because of their shape, these produce proteins that guide cell differentiation in advanced organisms).
- *VPR*: a gene governing reproduction of the HIV virus.

Further linkage studies identified genes for a variety of ailments including bipolar disorder, cerulean cataracts, melanoma, hearing loss, dyslexia, thyroid cancer, sudden infant death syndrome, prostate cancer, and dwarfism.

over who owns which parts of the genome. The scientists and research corporations have worked out a way to share access to a computerized database detailing 35,000 human genes. At the same time, gene therapy, immune system modulation, and genetically engineered antibodies entered the clinic in the war against cancer. Besides, in 1991, researches on human embryo were successful in inventing a technique for testing genetic abnormalities in an embryo. Meanwhile, genetic research resulted in the identification of an individual's genetic history and structure. The genetic structure of each human being is unique and cannot be compared with the genetic structure of another human being. Therefore, the identification of human beings on the basis of their genetic structure and identity would be the best possible way of identifying the individual. In 1992, the US Army begun collecting blood and tissue samples from all new recruits as part of a "genetic dog tag" program aimed at better identification of soldiers killed in combat. In the 1990s, biotechnology research and inventions led to the production of genetically engineered plants and animals on a large scale. These inventions could make genetically modified food with desired calories possible. However, there was an apprehension about the safety of these genetically modified foods for human consumption. Researches and trials on consuming such food could yield positive results, and FDA declared in 1993 that genetically engineered foods were "not inherently dangerous" and did not require special regulation. In the subsequent year, FDA gave market approval for the first genetically engineered food product, the Flavr Savr tomato, for human consumption. In the same year, unleashing the human genome and human chromosome came within the reach of guess work of the scientists when an international research team, led by Daniel Cohen, of the Center for the Study of Human Polymorphisms in Paris, produced a rough map of all 23 pairs of human chromosomes. When the entire world was guessing about research on cloning and genetically modified human beings, researchers at George Washington University cloned human embryos and nurtured them in a Petri dish for several days.[96] Further during the 1990s genome sequencing of a number of

---

[96] Although the project provoked protests from ethicists, politicians, and critics of genetic engineering, the research could take place and arrive at desired results.

organisms was done with success. These results encouraged the scientific community to work toward sequencing of the human genome at later stages,[97] a rough draft of which was ready by 1998. The draft showed the location of roughly 30,000 human genes. In 1997, the entire world was shocked with the birth of Dolly, a genetically engineered and cloned sheep. Researchers at Scotland's Roslin Institute reported that they had cloned a sheep named Dolly from the cell of an adult ewe. Subsequently, the researchers cloned another sheep Polly, the first sheep cloned by nuclear transfer technology bearing a human gene. In the following year, University of Hawaii scientists cloned three generations of mice from nucleus of adult ovarian cumulus cells. In 1998, the scientists at Japan's Kinki University cloned eight identical calves using cells from a single adult cow.

Biotechnology Meets Information Technology

We are living in an era where a multidisciplinary approach is very common. Everything, whether in profession or in occupation, has a multifaceted outlook. Research is no exemption. In particular, biotechnology is the result of a multidisciplinary approach and involves multifaceted techniques and methods that make it an absolute multidisciplinary technology. Biotechnology is also associated with information technology in its research and development.[98] In the recent past, these two technologies together gave rise to certain wonderful developments and milestones. Biotechnology and information technology came together in 1997 when using a bit of DNA and some biological laboratory techniques, the researchers engineered the first DNA computer "hardware."

---

[97] In 1996, a collaboration of scientists reported sequencing of the complete genome of a complex organism, *Saccharomyces cerevisiae*, otherwise known as Baker's yeast. The achievement marked the complete sequencing of the largest genome to date: more than 12 million base pairs of DNA. Further, in the same year, the sequencing of the genome of ancient organisms—archaea—found in inhospitable climates deep in thermal vents under the sea greatly advanced the understanding of the evolution of life on Earth. The microorganisms are neither eukaryotes nor prokaryotes. In 1998, the first complete animal genome of *Caenorhabditis elegans* was sequenced.

[98] See also Sreenivasulu. 2006. 'Information Technology and Biotechnology'.

Furthermore, a new DNA technique combined PCR, DNA chips, and computer programming, providing a new tool in the search for disease-causing genes. Further use of computers in research in these fields is continuously increasing. Computers and associated software are used in molecular modeling, rational drug design, design and production of compound libraries by combinatorial chemistry, performance and evaluation of high-throughput assays, and sequencing of DNA fragments and their correlation with known sequences. Inventions that relate to improvements in any of these techniques, in whole or in part, may be software related, and it is important to be aware of what may and may not be patented in this area.[99] Nowadays, drugs can be developed on a computer avoiding the inconvenience of using test tubes and chemicals for the physical design of the drug. Using a three-dimensional image of a protein structure, we can rotate the molecules on a screen and study them from all angles to figure out which drugs may best fit the active sites on a protein surface. Chemical engineers can manipulate different combinations of drugs and disease proteins. They can analyze the functional outcomes of adding one molecule from scratch using their computerized construction sets.[100]

The New Saga

During 1997–8, several therapies and treatments were invented and sought for approvals. For example, FDA granted marketing clearance to Remicade (infliximab), a novel monoclonal antibody for the treatment of Crohn's disease. Furthermore, an antibody therapy against breast cancer, Trastuzumab (Herceptin), heralded a new era of treatment based on molecular targeting of tumor cells. In 1998, genetic research exceeded all expectations with the discovery of stem cells that had the capacity to develop into any organ of the body or into complete living beings. These stem cells could be derived from bone marrow and from embryos.

---

[99] Grubb. 1999. *Patents for Chemicals, Pharmaceuticals and Biotechnology*, p. 261.

[100] Of course, the test question would be whether a drug acts only on the specific target and not on something similar, and whether it reacts in the body in the same way as it does on the computer screen or in a test tube.

The stem cells exhibit great ability in curing several diseases and hence are of immense medical value. However, stem cell research is associated with great ethical concerns, since it involves tinkering with a cell that has the capacity to grow into a complete living organism. However, with continued research, two research teams succeeded in growing embryonic stem cells, the long-sought grail of molecular biology. The high speed at which fragments of the genome could be sequenced led to the elucidation of the full DNA sequence of many human genes and partial sequencing of many more. This science of genomics can be used to find genes that could make useful protein products, which could be applicable in gene therapy and might be useful in the elucidation of disease mechanisms in diagnostic kits and screening for new drugs.[101] There are also many research tools and techniques that make use of biotechnological processes. The development of process engineering and protein engineering that involves protein production through rDNA technology is another achievement of modern biotechnology.[102] In fact, fermentation, pasteurization, sterilization, tissue culture, law of heredity, development of vaccines, discovery of antibiotics and their large-scale production, discovery of cell, cell culture, discovery of chromosomes, discovery of DNA and genes, isolation of genes, genetic engineering, gene slicing, gene mapping, gene therapy, hybridoma technology, raising of antibodies, discovery of the protozoan *Plasmodium* as a cause of malaria, invention of the technique of cloning, genetic modification of microorganisms, genetic modification of plants, genetic modification of animals, modification, isolation, and purification of human genetic material, in vitro fertilization, nonnatural insulin production, production of human growth hormone, hepatitis-B vaccine, investigation of anthrax, stem cell research, and embryonic research were all milestones in the development of biotechnology.[103]

[101] Grubb. 1999. *Patents for Chemicals, Pharmaceuticals and Biotechnology*, p. 225.

[102] Recombinant protein development involves isolation purification and in vitro production of intended protein. It requires the incorporation of a particular gene coding for particular protein into an intended host body (say bacteria) and expressing it.

[103] Smith. 1996. *Biotechnology*, 3rd edn.

Dominance over Nature

If it is accepted that biotechnology has its roots in history and has successful industrial applications, has there been public awareness on this subject in recent years? Undoubtedly, the main reason must derive from the rapid advances in molecular biology, in particular rDNA technology, which is allowing human beings dominance over nature. Using these new techniques it is possible to manipulate the hereditary material (DNA) of cells between different types of organisms, creating new combinations of characters and abilities not previously known to be present in the living organisms. The potential of these techniques first developed in academic laboratories is now being rapidly exploited in industry. The industrial benefits are immense, but the inherent dangers of tampering with nature must always be heeded and addressed. Undoubtedly, biotechnology has empowered human beings to have dominance over nature itself. Things that are natural can be altered by adding certain nonnatural characteristics, features, qualities, and traits through genetic engineering or rDNA technology.[104] This feature of biotechnology in changing living beings from their natural form to modified form certainly amounts to dominance over nature. Many such things that, at some point, seemed impossible or improbable have now become plausible with the help of biotechnology. Therefore, biotechnology has gained dominance over nature and it has empowered human beings to alter nature and natural things according to their thoughts and needs.

[104] See generally, Rockman. 2004. *Intellectual Property Law for Engineers and Scientists*.

2 Biotechnology Dynamics and Applications

Biotechnology inventions have impacted the lives of humankind throughout the world,[1] and examining this impact would enable us to identify and understand various legal and policy concerns related to it. Biotechnology has evolved through discovery of new techniques and methods that encouraged further research and development. An insight into and an understanding of several techniques of biotechnology will clearly mark out the scope and significance of biotechnology in the modern world. The effect, influence, and implications of biotechnology on the human life can be understood through its application in diverse fields, such as agriculture, medicine, animal husbandry, forestry, fisheries, protection of environment, and so forth. The legal concerns involved therein, including intellectual property claims, trade issues, environment concerns, public policy concerns, consumer concerns, and

[1] See Sreenivasulu, N.S. 2005. 'Biotechnology and Patent Law', *Journal on World Intellectual Property Rights*, 1(1–2, June–December): 213–34.

regulatory issues, arise from the use or development of these techniques and their application in diverse fields.[2] The following milestones in the history and development of biotechnology would precisely describe biotechnology, its armory, significance, capacity, and efficacy in meeting the different needs of society. This enquiry would enable us to understand better what biotechnology is,[3] what it consists of, what it can do, what it promises, and what it signifies. Let us discuss in detail the path-breaking inventions of biotechnology, ranging from ancient times to the modern world.

Microscope

The invention of microscope has changed the path of biotechnology. Without the microscope, the latest and modern techniques of biotechnology would not have been possible as it has facilitated most of the biotechnological research and discoveries. Although the invention of the microscope is related to the field of physics, it paved the way for the development of biotechnology and it has been responsible for several incredible discoveries that have permanently changed the modern world and its perceptions.

Plant Breeding and Hybridization

Plant breeding started with sedimentary agriculture and domestication of the plants by the early man. Soon, the search for superior plants to harvest began and the domestication was hastened by the early practice of harvesting mutant plants with special traits. This was the earliest style of plant breeding. Before Mendel's discovery, some plant breeding experiments, including selection and hybridization, had already been performed. However, plant breeding got a boost after Mendel's experiments on pea, leading to a new science called genetics. Modern plant

[2] See Sreenivasulu, N.S. 2005. 'Patenting Genetically Modified Life Forms: Legal Issues and Challenges', *Indian Bar Review*, XXXII(3&4): 4.

[3] See Sreenivasulu, N.S. 2006. 'Information Technology and Biotechnology', in T. Sabanna (ed.), *Globalisation and WTO*, pp. 205–16. New Delhi: Serials Publications.

breeding is applied genetics but its scientific basis is broader and uses conceptual and technical tools such as molecular biology, cytology, systematics, physiology, pathology, entomology, chemistry, and biometrics. Mendel's plant breeding experiments had stimulated research on plants with improved crop production.[4] The most famous contribution of his research was hybridization, which was responsible for a remarkable improvement in the economically important crops. Green revolution, which came into the forefront to fight hunger and food deficit and created food surplus, was the result of hybridization. Of the many crops developed during the green revolution, there were three very important crops, namely, hybrid maize, high-yielding wheat, and rice. We have experienced agrarian economies significantly benefiting from plant breeding and innovations in plant varieties.[5]

Tissue Culture and Cell Culture

Cell cultures are used to manufacture virus vaccines, to screen drugs for parasitological and toxicological effects in cancer and physiological research, and in the preparation of chemical substances. Tissue and organ cultures have also been used for similar purposes, which additionally could be utilized in grafts and organ transplants. Cells can be cultured to yield muscle, nerve, skin, tissue, and, in turn, organs. In plant breeding, it is an established practice to fuse two or more different cells into a single hybrid cell and cell lines. These hybrid cells contain genetic information from different cells, which has been transferred to the hybrid cell through culture. The technologies of "tissue culture" and "cell culture" technology essentially involve the production of cell lines. Cell lines are self-replicating; and in the development of living beings, body cell replication plays a vital role in shaping the body in a biologically required manner. Tissue culture involves producing cell lines outside the original host or the body of a living being and, in principle, these lines are immortal and constitute a constant supply of the relevant cells. The original host might be a bacterium, a plant or an animal, or even a human being. The use of human

---

[4] See Sreenivasulu. 2005. 'Biotechnology and Patent Law'.

[5] See Sreenivasulu, N.S. 2013. *Law Relating to Intellectual Property*, 1st edn, p. 381. IN, USA: Penguin-Partridge Publications.

genetic material, in particular, makes this technique controversial. Stem cell research is the most controversial of all. Stem cells are collected from the bone marrow, blood, or embryos and have the capacity to grow into a complete living being. In recent times, embryonic stem cells have ignited several controversies and complexities, since it involves collecting some tissue of cell line from a fetus or an embryo.[6]

Hybridoma Technology

The hybridoma technology deals with the human immune response and involves two of its key players: white blood cells and antibodies.[7] It yields hybridoma cells capable of manufacturing highly specific monoclonal antibodies. These cells are made by exposing spleen cells to an antigen, mixing the spleen's lymphocytes with cancer cells, adding a fusion initiator, and killing off the undesired cells.[8] At its most basic, antibody is produced through fusing white blood cells, which have the advantage of yielding production indefinitely, with tumour cells. When such a hybridized cell is injected into an antigen that is a foreign body, it would provoke an antibody response through which an indefinite supply of antibodies is ensured. This technique is invaluable for the research into the operation of the immune system, and has been used in the production of diagnostic and other testing kits.[9]

Genetic Engineering/Recombinant Deoxyribonucleic Acid (rDNA) Technology

Genetic engineering is probably the most well-known or at least the most popular technique in biotechnology. Genetic engineering is also

---

[6] See Plomer, A., P. Torremans, B. Knoppers, C. Denning, J. Sinden, and M. Levin. 2006. 'Stem Cell Patents: European Patent Law and Ethics Reports', Report for the European Commission, 28 EIPR 11, 569–75.

[7] Torremans, P. 2008. *Holyoak & Torreman's Intellectual Property Law*, p. 85. New York: Oxford University Press.

[8] Cooper, I.P. 1999. *Biotechnology and the Law*, pp. 1–14. New York: West Group.

[9] Overwalle, G.V. (ed.). 2007. *Gene Patents and Public Health*. Brussels: Bruylant.

known as recombinant DNA (rDNA) technology. It manipulates living matter at a subcellular level that involves manipulation of cells, which in turn means manipulation of the genome of the cell. The genome of the cell contains chromosomes, and chromosomes contain DNA. The actual manipulation takes place at the DNA level. Genetic engineering is the use of cloning vehicles or vectors to introduce foreign DNA into host cells under circumstances guaranteeing that the host cells will cause that foreign DNA to replicate and that the genes carried by the foreign DNA will be able to express themselves.[10] Genes are placed on certain portions of DNA, which makes the DNA functional in dictating functions to the cell.[11] The active portions/functional parts of DNA are known as genes, which activate the production of protein, enzyme, or any other genetic material in accordance with the function of the respective cell that contains them.

Genes are made up of amino acid sequences, which combine to produce proteins. The release of proteins marks the significance of the gene and makes the cell function in different ways. Genetic engineering involves incorporation of new genes into a cell or removal of a functional gene from a cell to affect different kinds of function to be performed by the cell. If a new gene is incorporated or a gene is removed, it affects the release or nonrelease of a specific protein, which in turn affects the functioning of a cell. Therefore, genetic engineering is done with an intention either to stop a gene from being instrumental in releasing a protein or to affect the release of a new protein in the cell to influence the actual functioning of the cell altogether. Genetic engineering involves a few steps. The first step would be to identify the gene, and the second step would be to isolate the intended gene. The third step would be to purify the gene, and the fourth step would be to incorporate the same into the DNA of an intended cell.[12]

[10] Cooper. 1999. *Biotechnology and the Law*, pp. 1–9.

[11] Grubb, P.W. 1999. *Patents for Chemicals, Pharmaceuticals and Biotechnology: Fundamental of Global Law, Practice and Strategy*, p. 230. Oxford, New York: Clarendon Press.

[12] See generally, Kanakala, K.C. 2007. *Genetic Patent Law and Strategy*, 1st edn, pp. 164–5. Noida: Manupatra Publications.

Antisense Technology

It is a technique through which the expression or function of a gene can be suppressed or nullified. This technology will prevent a gene that could be responsible for a specific disease in a living organism from releasing the protein or enzyme that causes such a disease.[13] It involves binding the gene with a DNA strand and blocking transcription of that particular gene.[14] Gene therapy functions on the basis of this technology to cure hereditary diseases.

Genetically Modified Organism

It is an organism whose genetic material is altered through genetic engineering so that the organism possesses certain nonnatural, genetically modified physical and chemical properties, qualities, and features that make it unique and special. A genetically modified organism could be a microorganism such as a bacterium, or it could also be a plant or an animal. In 1973, the first-ever genetically modified organism was created. In 1980, the first-ever genetically modified microorganism was patented. In 1985, a genetically modified plant and, in 1988, a genetically modified animal were patented, followed by patenting of modified, isolated, and purified human genetic material. Nowadays, genetic modification and genetic engineering have become words of common usage. A genetically modified organism may be identified by virtue of its new phenotypic characteristics attributable to the composite plasmid.[15] Genetic modification of an organism is a herculean task and involves isolation and purification of genetic material of one organism and incorporating it into another organism. In turn, such an organism would become genetically modified with foreign genetic material in its genome, to see it perform certain nonnatural function for a desired result.

---

[13] Grubb. 1999. *Patents for Chemicals, Pharmaceuticals and Biotechnology*, p. 242.

[14] See Grace, E. 1998. *Biotechnology Unzipped: Promises and Realities*, pp. 74–5. Hyderabad, India: Universities Press.

[15] Cooper. 1999. *Biotechnology and the Law*, pp. 1–9.

Cloning

Cloning involves the creation of a genetically identical organism. It is a technique whereby the nucleus of a cell from an intended organism whose replica has to be produced would be incorporated into the embryo in place of the nucleus of the embryo in the mother's womb.[16] Once the embryo is altered, the same shall be implanted into the mother's womb, wherein it grows and takes birth as the replica of the intended organism. Cloning is perhaps the most controversial of all biotechnological innovations, and legal regulation has been strict in this regard. Cloning can primarily be distinguished on two grounds: the procedure or the technique used for cloning and the application or the purpose for which cloning is done.[17]

1. Based on the procedure or the technique used, cloning is of two types:
   • Blastomere separation technique; and
   • Somatic cell nuclear transfer (SCNT) technique
2. Based on the application or purpose of cloning, cloning is further classified as follows:
   • Reproductive cloning, that is, cloning with the intention of creating human babies; and
   • Therapeutic cloning or non-reproductive cloning, that is, cloning with the intention of harvesting stem cells from the embryo.

The blastomere separation technique or the embryo splitting technique is a simpler procedure adopted for cloning, as it does not involve nuclear transfer of somatic cells. This technique deals with the splitting of an early embryo, allowing each split blastomere cell to grow into a separate individual organism. Therefore, this process of cloning is

[16] See generally, Kanakala. 2007. *Genetic Patent law and Strategy*, p. 172.

[17] Sreenivasulu, N.S., Rohan Benerjee, and Arpan Narayan Choudary. 2016. 'Cloning Technology, Public Policy and Human Rights', in N.S. Sreenivasulu (ed.), *Human Rights and Development*, pp. 155–75. Bloomington, Indiana, USA: Penguin-Partridge Publications.

adopted to replicate only a fertilized egg and is not applicable for clon-ing adult cells.[18] However, cloning by SCNT is a more complex and sophisticated process that involves the removal of a haploid nucleus from an unfertilized egg cell and its replacement with a diploid nucleus from the donor somatic cell. The somatic nucleus in the "reconstructed" egg cell reprogrammed by the components of the egg cell along with artificial stimuli begins to divide and develops into an embryo (embryogenesis).[19] The technique of SCNT has a long history dating back to the 1960s and it was adopted for cloning of a sheep, Dolly, from an adult somatic cell.[20] It is the most commonly preferred procedure to create clones, since it can also duplicate an adult cell. There is no requirement of a fertilized egg in this technique, and a clone can be created from a somatic cell itself.

Human Genome Project

A different way of looking at human genes is by sequencing all or part of the entire human genome, finding which sequences correspond to expressed genes, correlating gene expression with cell type and disease state, and gradually building up a picture on the function of the genes that are found.[21] With this basic approach, the National Institute of Health set up the Human Genome Project under the Human Genome

[18] In fact, this process closely resembles a natural biological situation where a zygote is split to create twins. The blastomere separation technique has also passed scientific scrutiny in production of mammals when, in 1995, Rosalin Institute in Edinburgh created the two cloned sheep, Morag and Megan, from a single differentiated embryo. However, this process is considered a distant second substitute to the SCNT process. See http://cosmos.ucdavis.edu/archives/2007/cluster7/seraphin_rebecca.pdf (last accessed 1 February 2016).

[19] See Shalev, C. 2002. 'Human Cloning and Human Rights: A Comm-entary', *Health and Human Rights*, 6(1): 137–51.

[20] In fact, Gurdon wrote about this process in 1962; see Burley, J. 2007. 'An Abstract Approach to the Regulation of Human Genetics: Law, Morality and Social Policy', in H. Somsen (ed.), *The Regulatory Challenge of Biotechnology: Human Genetics, Food and Patents*, pp. 83–4. Cheltenham, England: Edward Elgar Publication Ltd.

[21] Grubb. 1999. *Patents for Chemicals, Pharmaceuticals and Biotechnology*, p. 242.

Organization (HUGO) initiative in Washington, US, in 1988. This project was undertaken to unravel the secrets of life, the aim being to sequence the entire genome of a human being to find out the function of every respective gene. Recently concluded, the project revealed the functions of different genes inside the genome. This project has helped us to understand the protein or enzyme codes and how these can be used for producing medicines or for preventing and curing diseases. It has initiated the debate on human genome being the heritage of humanity.[22]

Penicillin

In 1928, Alexander Fleming invented penicillin. He observed that colonies of the bacterium *Staphylococcus aureus* could be destroyed by the mold *Penicillium notatum*, proving that there was an antibacterial agent. It led to the invention of medicines that could kill certain types of disease-causing bacteria inside the body. At the time, however, the importance of penicillin was not known. In fact, the use of penicillin began in the 1940s when Howard Florey and Ernst Chain isolated the active ingredient and developed a powdery form of the medicine, opening the road to antibiotics.

Antibiotics

Today, antibiotics are responsible for curing several diseases, providing much required resistance power to the otherwise diseased body. Biotechnology provides the ways and means of identification and isolation of natural antibodies produced by the living beings that are used for preparing antibiotics. The immune systems of living beings produce antibodies inside the body naturally, which are supposed to release enzymes to resist or fight against diseases. Research has made it possible for natural antibiotics to be produced in vitro, that is, outside the body, for commercial uses. Antibiotics should be used only under medical prescription and their overdose needs to be avoided. The more the antibiotics are used, the more resistant the surviving strains become.

---

[22] Francioni, F. and T. Scovazzi (eds). 2006. *Biotechnology and International Law*, p. 302. Portland, USA: Hard Publishing.

Those strains that survive an antibiotic assault that kills 99.9 percent of their fellow germs are more likely to be superstrains, almost invincible to control for a long time to come.[23]

Vaccines

Vaccine comes from the Latin word "vaccines" meaning "from cows." The invention and use of vaccines can be traced back to smallpox, a disease that threatened the entire humankind in the eighteenth century. There was no medicine or drug available at that point that could cure the disease. The first vaccination was developed in 1798 when Edward Jenner used cowpox virus to heal the smallpox virus.[24] Jenner, with an ambition to find a vaccine against smallpox, studied the relation between smallpox and, yet another disease, cowpox, and observed that cowpox virus could be used in preventing smallpox. He was successful in curing smallpox after injecting cowpox virus into the body of a boy. In 1798, he published his study, which became quite popular across the world. Smallpox has been completely eradicated now using the method invented by Edward Jenner. It was followed by the invention of vaccine for polio and such other diseases. The world is still fighting against polio and there is continued research for finding vaccine for several other diseases threatening the modern world. Vaccines are the second largest category of over 200 drugs being produced by American pharmaceutical companies using biotechnology.[25]

Insulin

Insulin is yet another breakthrough in the history of biotechnological research. Production of insulin through biotechnology for human use has been a major milestone in the biomedical field. The first biotechnological drug to be approved and marketed was perhaps "humulin," which is insulin. Humulin is produced by a genetically modified bacterium

---

[23] Grace, E. 1998. *Biotechnology Unzipped: Promises and Realities*, p. 92. Hyderabad, India: Universities Press.

[24] Cooper. 1999. *Biotechnology and the Law*, pp. 1–13.

[25] Grace. 1998. *Biotechnology Unzipped*, p. 81.

that resides in the human digestive tract to create a product identical to human insulin.[26] Prior to this development, insulin was obtained from the pancreas of domesticated animals.

Organ Transplantation

In case of damage or failure of any organ of a living being, biotechnology helps in the culture of new organs and its transfer into the body in need. The culture or creation of a new organism is possible through cell culture or tissue culture. Recent researches have been successful in the culture and production of organs for the need of living beings. There exist several apocryphal accounts of transplants much prior to the existence of necessary scientific understanding and advancements. The Chinese physician Pien Chiao reportedly exchanged hearts surgically between a man of strong spirit but weak will and a man of weak spirit but strong will in an attempt to achieve balance in each man. Roman Catholic accounts report the third-century saints Damian and Cosmas as replacing the gangrenous leg of the Roman deacon Jastinian with the leg of a recently deceased Ethiopian. Most accounts have the saints performing the transplant in the Fourth Century and later. The more likely accounts of early transplants deal with skin transplantation. The first reasonable account is of the Indian surgeon Sushruta in the second century BC, who used autografted skin transplantation in nose reconstruction rhinoplasty. Joseph Murray performed the first successful transplant, a kidney transplant between identical twins in 1954.[27] In fact, it was the advent of cyclosporine that altered transplants from research surgery to life-saving treatment. In 1968, surgical pioneer Denton Cooley performed 17 transplants including the first heart–lung transplant, which was not successful. Fourteen of his patients died within six months after transplant. By 1968, two-thirds of all heart transplant patients survived for five years or more. With organ transplants becoming common, researchers started working on more complicated multiple-organ transplants

---

[26] Naidu, D. 2009. *Biotechnology and Nanotechnology: Regulation under Environmental, Health and Safety Laws*, p. 183. London: Oxford University Press.

[27] No immunosuppression is necessary in genetically identical twins.

on humans and whole body transplants on animals. In 1981, the first successful heart–lung transplant took place at Stanford University Hospital. These organ transplantations took place with the organs produced through cell culture, tissue culture, and organ cultures.

Transgenic Microorganisms

A microorganism can only be seen through a microscope and whose genome can be altered to get the desired results. The study of microorganisms involves the study of algae, fungi, bacteria, protozoa, and viruses, which otherwise are known as lower life forms. These microorganisms, which are found naturally in soil, water, plants, animals, sewage, and the like, can be genetically modified to incorporate special traits and features for a desired purpose. Such modified organisms are called transgenic microorganisms. In 1980, the first-ever transgenic microorganism was patented. A scientist, Anandha Chakraborty, had developed a genetically modified bacterium that had the capacity to eat and clean crude oil spills.[28] This ability was installed into the bacterium through genetic engineering.

Transgenic Plants

The genetic setup of a transgenic plant is altered through genetic engineering. Such alteration can be done either by removing a gene or by incorporating a gene, or by suppressing a gene and other genomic material inside the cell. On completion of the transformation, conventional breeding techniques enable the production of plants whose seeds will pass the desired features to next generations.[29] For example, in the case of Bt cotton, the genetic structure of the cotton plant is altered to give it certain foreign genes, which provide certain novel characteristics such as high yield, quick growth, little water consumption, tolerance for drought conditions, and so forth. The long-term aim is to generate transformed plants that would have not only the desirable characteristics that are

[28] See Sreenivasulu. 2006. 'Information Technology and Biotechnology'.

[29] Grubb. 1999. *Patents for Chemicals, Pharmaceuticals and Biotechnology*, p. 251.

normally attainable by conventional breeding programs but all the extra advantages such as high yield, growth in arid conditions, additional nutritional quality, and others that would be necessary[30] in order to be able to meet the growing demand for food. With the population estimated to exceed ten billion by 2050, and considering that there would be less arable cultivating land at that point than at present mainly because of industrialization, transgenic plants capable of giving better yield with various advantages could prove crucial.

Transgenic Animals

The genome of a transgenic animal is altered to provide it with certain features and traits that are unique to it. Various techniques, including microinjection, make it possible to introduce extraneous genetic material into a fertilized mammalian ovum, insert the ovum into a pseudopregnant female, and obtain offspring in which the intended genetic material is incorporated into the genome. By combining this process with classical breeding steps such as backcrossing, it is possible to obtain an animal strain that stably transmits the new gene to subsequent generations, which will display the corresponding phenotype.

Transgenic animals have three different but important utilities at present. First, they can be used as animal models for research. For example, Harvard University has developed a genetically engineered mouse that is susceptible to cancer.[31] Through genetic engineering, a gene responsible for cancer was injected into the genome of the mouse at the stage of fertilization in the mother's womb. The fertilized egg, when it grew, possessed the gene responsible for cancer and genetically became susceptible to cancer. This mouse, which is genetically susceptible to cancer, is used as an animal model in research, in particular to test cancer-treating drugs and medicines.[32] Second, transgenic animals can give a high yield of flesh, milk, and wool with imbedded capacity to resist and withstand diseases; for example, genetically engineered cow

---

[30] Grubb. 1999. *Patents for Chemicals, Pharmaceuticals and Biotechnology*, p. 251.

[31] See Sreenivasulu, N.S. 2013. *Law Relating to Intellectual Property*, p. 352.

[32] See Sreenivasulu. 2005. 'Biotechnology and Patent Law'.

capable of giving high yield of milk, genetically engineered sheep capable of giving high yield of wool, and so on. Third, a transgenic animal is a source of organ transplantation. For example, pig's organs are being used as replacement for human organs. The research results indicate that although natural pig organs may or may not suit the need, yet genetically engineered pig organs can be of significant use for organ transplantation.

Isolated and Purified Human Genetic Material

Human genetic material, which is available in the natural, unisolated, and unpurified form in living beings, can be identified, isolated, and purified to incorporate it into some other organism for commercial purposes. Isolation of human genetic material such as DNA and genes can be done through genetic engineering. Human genetic material that codes for some proteins or enzymes having medicinal values can be produced on commercial basis after isolating it from the human body, which is its natural source, and incorporating it into other organisms such as bacterium.

Gene Therapy

Genes are responsible for genetic or hereditary diseases. Since genes are passed over generations, they carry the genetic information of the parents to the children. If a parent is suffering from a genetic disease, it gets transferred to the children.[33] If a particular gene responsible for a disease could be removed from the body of a living being, it would make the present and all future generations free from that genetic disease. This process of removing the disease-causing gene is known as gene therapy. It is also a technique to correct defective genes responsible for the disease. Through gene therapy, a desired gene can also be inserted into an intended organism. *Altered Fates*, written by two prize-winning journalists, Jeff Lyon and Peter Gorner, gives an engaging account of the story of the human race to carry out gene therapy.[34] In fact, tampering

[33] See generally Kanakala. 2007. *Genetic Patent Law and Strategy*, p. 170.

[34] For details on gene therapy, how it evolved, and story behind it, see Grace E. 1998. *Biotechnology Unzipped: Promises and Realities*, pp. 64–9. Hyderabad, India: Universities Press.

with human genes, and removing genes from or incorporating them into human beings, started only after these had been successfully researched and tested on plants and animals for more than 20 years. The deliberate insertion of a gene into a human body happened about 20 years after the researchers were able to transform and recombine genes in bacteria, fruit flies, tomatoes, toads, mice, and other species.[35] The first actual use of gene therapy began in September 1990, when a child suffering from a rare genetic immunodeficiency disease was treated.

Stem Cell Research and Embryonic Research

Stem cells have the capacity to develop into a complete organ, living being, or human being. They are derived from the bone marrow, blood, and embryos. In case of organ failures, stem cell research helps in developing required organs to be transplanted into the body of the patient. A complete organ or a living being can be developed using stem cells collected from the embryo, blood, or the bone marrow. Furthermore, embryonic research involves using embryos for medicinal purposes such as developing drugs, vaccines, medicines, as well for required organs. Both stem cell research and embryonic research are controversial, as they use embryos, which can develop into complete human beings, for research.

Business Applications of Biotechnology

Hindu mythological stories and epics tell tales of God's birth on the earth in different times and taking different forms. In particular, the Mahabharata makes a mention of Lord Krishna taking 10 different avatars (characters) with 10 faces or heads that he exhibited at times. Similarly, another epic, the Ramayana describes Ravana having 10 heads. In the contemporary times, biotechnology is poised to have more than 10 faces in terms of its nature, implications, and applications. The

---

[35] Although humans are not different from other forms of life as far as DNA is concerned, it took more than two decades to test gene transformation and gene therapy on humans after their success with other life forms for various reasons.

successful application of biotechnology depends upon the cooperation of the experts in different fields. The scope and applications of biotechnology are increasing day by day.[36] At one point of time, biotechnology was limited to the fermentation of beer, but today it has played significant role in revolutionizing[37] certain industries such as agriculture, animal husbandry, medical and pharmaceutical, forestry, fisheries, environment protection, chemicals, food, beverages, etc.[38] The applications of biotechnology are so broad, and the advantages so compelling, that virtually every industry is using it. There are developments in areas as diverse as pharmaceuticals, diagnostics, textiles, aquaculture, forestry, chemicals, household products, environmental cleanup, food processing, and forensics to name a few. Biotechnology is enabling these industries to make new or better products, often with greater speed, efficiency, and flexibility. Business application or commercialization or industrialization of biotechnology has been on the go in the recent past. Industrialization of biotechnology involves the application of different techniques of modern molecular biology to improve the efficiency and reduce the environmental impacts of the processes of industries such as textile, paper and pulp, and chemical manufacturing, to name a few. For example, industrial biotechnology companies develop biocatalysts such as enzymes to synthesize chemicals. Enzymes are proteins produced by all organisms. Using biotechnology, the desired enzyme can be manufactured in mass commercial quantities. A wider application of biotechnology for industrial or business purposes at a global level will depend on the strengthening of an international governance system for the new bio-economy that is based on the principles of market inclusion.[39] Biotechnology industry is growing and active, raising concerns of intellectual property, trade, environment, consumer interest, GM food, public policy, etc. The active areas within this industry include medical, pharmaceutical and

[36] In fact, biotechnology seems to be a resort to the different problems existing in the society, such as poverty, diseases, pollution, and drought.

[37] Watal, J. 2002. *Intellectual Property Rights under WTO and Developing Countries*, p. 128. Oxford, UK: Oxford University Press.

[38] Sreenivasulu. 2006. 'Information Technology and Biotechnology'.

[39] Ricardo Melendez, O. and V. Sanchez. 2007. *Trading in Genes: Development Perspectives on Biotechnology, Trade and Sustainability*, p. 11. London: Earth Scan.

cosmetics, agriculture, consumable foods, environment and sustainable development, livestock, animal husbandry, and energy to name a few where biotechnology is being applied.[40] Let us discuss in detail each of these active areas and applications of biotechnology where potential law and policy issues as contemplated need to be addressed.

Medical, Pharmaceutical, and Cosmetics

Biotechnological methods are now used to produce proteins for pharmaceutical and other specialized purposes. Biotechnology represents a powerful alternative to traditional methods of drug discovery that involve laborious screening of thousands of organic compounds found naturally in soil plants and molds.[41] The advent of rDNA technology or genetic engineering has paved way for the invention of innovative medicines and drugs possessing natural antibodies and proteins. Biotechnology has revolutionized the medical and pharmaceutical industry with its innovative way of producing medicines. Multinational biotechnology companies such as Novartis, Genentech, Biogen, Ciba Gigzy, Bio-con, and Dr. Reddy's Laboratories have been engaged in promotion of the application of biotechnology in the medical and pharmaceutical fields. The combination of biotechnological and traditional methods of producing medicines is yielding outstanding results.[42] It has provided cures and therapies for many diseases. Researches are being undertaken to find vaccines against typhoid, cholera, AIDS, and cancer. Recently Novavax, an American biotech company, tied up with Bharath Biotech Company based in Hyderabad for producing vaccines for different diseases including the recent outburst of "bird flu."[43] The role of biotechnology in producing antibodies and vaccines production

[40] Epstein, M.A. 2008. *Epstein on Intellectual Property*, 5th edn, pp. 2–3. New Delhi: Wolters Kluwer (India) Pvt Ltd.

[41] Dryfuss, R., D.L. Zimmeman, H. First. *Expanding the Boundaries of Intellectual Property: Innovation Policy for the Knowledge Society*, p. 257. New York: Oxford University Press.

[42] Smith J.E. 1996. *Biotechnology*, 3rd edn, pp. 129–31. Cambridge, UK: Cambridge University Press.

[43] *The Hindu Daily*. 2006. 'Vaccines for an Influenza Pandemic', 18 March, p. 11.

and in controlling diseases is indispensable. The diseases once considered incurable are now being cured through therapies invented by biotechnology. Recent applications of biotechnology in the field of medicine have given us major insights into diabetes, cancer, and other metabolic diseases.[44] In 1982, genetically engineered insulin got market approval in the UK and the US. In the same way, in the mid-1980s, genetically engineered growth hormone got approval. Furthermore, genetically engineered proteins to treat heart attacks and strokes, new vaccines for foot and mouth disease, and monoclonal antibodies to boost body's defense system against cancer and other diseases were introduced. In 1986, a vaccine derived from genetically engineered yeast was made available for hepatitis B, a major public health problem threatening the entire world.[45] Biotechnology is being used to manufacture products that are intended for therapeutic, preventive, and diagnostic purposes.[46] The application of biotechnology in the field of medicine has led to the development of "gene therapy." New genetic therapies are being developed to treat diseases such as cystic fibrosis, AIDS, and cancer. Applications of biotechnology in medical and pharmaceutical sectors include the development and production of human healthcare products, including drugs, vaccines, diagnostics, and therapies. It also involves the development of pharmaceutical products for the cure or control of human diseases, including enzymes, antibiotics, vaccines, gene therapy, and pharmaceutical, cosmetic, and household products. The major pharmaceutical companies now increasingly support biotechnology by either acquiring it from biotech ventures or by developing it in-house. Biopharmaceuticals have become the focus for a biotechnology-driven industry.[47] Furthermore, in the field of cosmetics, bio-cosmetics based on innovative researches in biotechnology have been introduced. Developing skin, tissue, and cells for cosmetic surgeries has become the order of the day. Doctored bodies and lifestyles

[44] Kornberg, A. 1995. *The Golden Helix: Inside Biotech Ventures*, p. 9. CA, USA: University Science Books.

[45] Ranga, M. 1999–2000. *Animal Biotechnology*, p. 4–10. Jodhpur, India: Agro Bios India.

[46] Naidu. 2009. *Biotechnology and Nanotechnology*, p. 7.

[47] Kornberg. 1995. *The Golden Helix: Inside Biotech Ventures*, p. 14.

have become a reality today with the help of biomedical research. Bio-cosmetic therapies, surgeries, and treatments to enhance one's beauty or looks are promising.

Agriculture, Food, and Consumables

Using new and promising scientific processes and techniques of biotechnology to crop germplasm and to crossbreed or mutate plants, animals, and microorganisms for improving crops and livestock are yielding good results. The scientific community and industry leaders are quite enthusiastic about the benefits of biotechnology for agriculture and consumable food sector, especially in terms of increased yield, resistance to diseases, pests, and weeds, increased food production, enhanced nutrition, and environmental protection.[48] Biotechnology has resolved the food requirements of the society by guaranteeing the right to food.[49] The green revolution in India played an important role in achieving self-sufficiency in foodgrain production, along with infrastructural development. The relevance of agricultural biotechnology is a result of the need for infusing technological change in the Indian agriculture sector to feed India's growing population. However, public opinion across the world about all forms of biotechnology is divided, and India is no exception. The emergence of agricultural biotechnology has created a serious debate in the country on its potential to revive Indian agriculture. Revolutionary changes in the field hold the promise of good yields achieved through the application of biotechnology in the agricultural sector.[50] In the era of biotechnology, methods such as genetic engineering are replacing traditional methods such as crossing, budding, and tissue culture to increase production. Applications of biotechnology in the sectors of agriculture and consumable food sectors involve agricultural/veterinary area product development, including plant and animal cultures for increased food production and food security, animal vaccines, and drugs and diagnostics for animal and plant development and sustainability. Multinational

---

[48] Francioni, F. and T. Scovazzi (eds). 2006. *Biotechnology and International Law*, p. 259. Portland, USA: Hard Publishing.

[49] Ranga. 1999–2000. *Animal Biotechnology*, p. 3.

[50] Smith. 1996. *Biotechnology*, p. 162.

biotechnology companies such as Monsanto, Rice Tech, and the like are very active in the agricultural biotechnology field. Use of biotechnology and genetic engineering in the field of agriculture has mainly focused on breeding plants that are resistant to herbicides or insects, such as genetically modified corn, soy, and cotton. However, plants can also be genetically engineered to make them drought resistant, more nutritious, or capable of being grown under conditions of salinity.[51] According to biotechnologists, the only way to provide food security is to genetically engineer crops. Biotechnology has introduced genetically modified food (GM food) to the world, which has revolutionized the food industry.[52] There is evidence to suggest that GM foods have high levels of proteins and vitamins compared with regular food, and they hold the promise of improved quality, nutrition, safety, and long-term preservation.[53] There is development of fermented products, production of chemicals used in food by microorganisms, and production of a wide range of food products, beverages, and ingredients. Foods developed through biotechnology have already been part of consumers' daily diet. Application of biotechnology in the food industry has influenced the taste, color, consistency, quality, preservation, health benefits, and cost of food. Plants such as rice, wheat, tomato, potato, and the like are genetically engineered to increase the consumable value.[54] A Chinese study demonstrated that a 10 percent increase in yield was witnessed with insect-resistant GM rice. For instance, GM rice and wheat have higher yields compared to regular rice and wheat. Furthermore, GM tomato or potato possesses longer shelf-life and gives high yield compared to regular tomato or potato.[55] These food items also possess additional traits such as more perseverance, good quality, and consumer liking.

[51] Naidu. 2009. *Biotechnology and Nanotechnology*, p. 7.

[52] Shahalia, M.L. 2003. *Perspectives in Intellectual Property Law, Many Sides to a Coin*, pp. 128–51. Delhi: Universal Law Publishing Company Pvt Ltd.

[53] Arup, C. 2000. *The New World Trade Organizations Agreements*, p. 216. UK: Cambridge University Press.

[54] See generally, *The Economist*. 1999. 'Genetically Modified Foods: Food for Thought', 17 June, pp. 23–7.

[55] Sreenivasulu, N.S. 2008. *Biotechnology and Patent Law: Patenting Living Beings*, 1st edn, p. 34. Noida: Manupatra Publications.

Environment Protection and Sustainable Development

Applications of biotechnology in the sphere of environment protection and sustainable development include developing specific microorganisms for breaking down or cleaning up oil spills and pollutants, biosafety, waste treatment, bioremediation, and environmental sustainability.[56] Pollution control is a matter of concern all over the world and various efforts are being made to preserve and protect the environment from being polluted.[57] Pollution can have immediate catastrophic effects on birds, animals, and the balance of ecology. The main causes of environment pollution are oil-derived compounds, toxic chemicals, industrial wastes, sewage, domestic wastes, gases, and release of fumes by vehicles. Environmental biotechnology plays a significant role in the protection of environment and helps in waste treatment and pollution control. It can clean up wastes more efficiently than conventional methods and greatly reduce our dependence on land-based disposal. In fact, natural microorganisms contribute significantly to the treatment of waste, sewage, pollutants, and other contaminated materials. Treatment of waste and sewage also prevents spreading of infectious diseases.[58] Furthermore, composting is also a biotechnological process that grows microorganisms to decompose solid organic wastes into a stable form considerably reduced in bulk, which can be safely returned to the environment. Biotechnology helped in discovering different processes to treat wastes and modifying the existing pollution treatment processes to yield better results. The decision of the Supreme Court of America in *Diamond v. Ananda Chakrabarty*[59] projects how biotechnology could be used to clean up pollutants and wastes. Biotechnology promises a clean and green environment through its innovative methods of pollution control and waste treatment. Therefore, the application of biotechnology to protect the environment is becoming necessary and inevitable due to ever increasing pollution.

[56] See generally Grace, E.S. 1997. *Biotechnology Unzipped: Promises and Realities*, pp. 133–4. Hyderabad, India: Universities Press.

[57] Sreenivasulu. 2005. 'Biotechnology and Patent Law'.

[58] Smith. 1996. *Biotechnology*, p. 145.

[59] 447 US 303 (1980).

Livestock and Animal Farming

In the field of livestock, the application of biotechnology through tissue culture or genetic engineering aims to produce a high yield of milk, meat, and wool. Tissue culture involves isolating and culturing cells or organs of livestock in vitro for a desired result,[60] and it is useful in testing drugs and medicines. We have seen hybrid animals produced by crossing two different animals with different genetic setup. Crossing a disease-resistant animal with a high milk-yielding animal may result in the offspring possessing both the desired qualities.[61] Genetic engineering now has greater potential to achieve desired results in animal husbandry.[62] Genetic engineering of animals involves isolation of specific genes, coding for certain characteristics, and incorporation and expression of the same in an intended animal to get desired results. Livestock is used in testing drugs, medicines, and therapies before releasing them for human use. Sometimes genetically engineered animals are produced solely for the purpose of testing drugs and medicines. Furthermore, certain proteins, antibodies, and enzymes produced inside the body of different animals have medicinal values. Such proteins, if produced commercially, can be used to prepare drugs and medicines. The body of a pig is perhaps called a pharmaceutical factory.[63] It produces many proteins and antibodies that have significant medicinal values. Biotechnology promises commercial production of these proteins through isolation and incorporation of the specific gene coding for the proteins into an intended organism. Such a genetically engineered organism produces the desired proteins in its body. Bacteria multiply rapidly and can produce proteins in a short period. Therefore, generally bacteria are used to produce proteins of different organisms (animal, plant, and human being) by incorporating the respective gene from the respective living beings. The vaccine for

---

[60] Sreenivasulu, N.S. 2008. *Biotechnology and Patent Law: Patenting Living Beings*, p. 26. Noida: Manupatra Publications.

[61] See generally Grace, E.S. 1997. *Biotechnology Unzipped: Promises and Realities*, p. 1000. Hyderabad, India: University Press.

[62] Department of Biotechnology. 1995–6. *Annual Report*, Government of India.

[63] Sreenivasulu. 2005. 'Biotechnology and Patent Law'.

Hepatitis B virus is developed from genetically engineered yeast in the aforementioned manner.[64]

Energy Sector

Biotechnology is used in the energy sector for the production of energy, biogas, and biofuels. Through the production of energy crops and biofuels, some of the serious problems of high prices and possible disruption of the supply in international petroleum markets that most countries are facing at present could be resolved. It may facilitate access to energy, especially for poor people in the developing countries; contribute to environmental preservation by reducing greenhouse gas emissions; provide new opportunities to rural communities in the developing countries; and add new value to agricultural commodities. The application of biotechnology in the energy sector may serve as a means for the developing countries to benefit from sustainable investments deriving from carbon trading activities.[65]

Varied Applications of Biotechnology

Biotechnology has impacted many fields in different ways. In the field of equipment and energy, biotechnology is used in the production of hardware, bioreactors, software, biochips, and so on. Biotechnology plays a vital role in extracting metals in the mining industry.[66] In the chemical industry, biotechnology is being used to develop or split chemicals. In forestry, biotechnological techniques such as tissue culture, genetic engineering, and cloning are used to conserve and preserve species that are on the verge of extinction. In the field of aquaculture, biotechnology is used to increase fish production, as fish cultivation can be improved through gene transfer and cloning. As the genes are transferred from one

---

[64] Ranga. 1999–2000. *Animal Biotechnology*, p. 6. The vaccine has become available in 1986.

[65] See Wuger, D. and T. Cottier (eds). 2008. *Genetic Engineering and the World Trade System: World Trade Forum*, p. 173. UK: Cambridge University Press.

[66] Smith. 1996. *Biotechnology*, p. 157.

animal or plant to another, they can also be transferred to fishes to get better results such as high protein values, disease resistance, production of nutritional or medicinal proteins, and high yield.

A Multidisciplinary Approach of Biotechnology

Biotechnology is a discipline of science that combines several other disciplines in its fold. In fact, it is the most diversified form of science exhibiting a bewildering array of subdisciplines such as fermentation technology, microbiology, biochemistry, protein chemistry, protein engineering, chemical engineering, immunology, enzymology, cell biology, cell culture, tissue culture, molecular biology, genetics, computer science, and so on.[67] In the recent decades, a characteristic feature of the development of science and technology has been the increasing resort to multidisciplinary strategies for the solution to various problems. This has led to the emergence of new interdisciplinary areas of study with the eventual crystallization of new disciplines with identifiable characteristics concepts and methodologies. The World Intellectual Property Organization (WIPO) study on the importance of Biotechnology for National Growth and Development stresses[68] the significance of biotechnology in the modern world.[69] If a field is multidisciplinary, it means that it has got quantitative extension of approaches to problems that commonly occur within a given area. It involves marshalling of concepts and methodologies from a number of disciplines and applying them to a specific problem in another area.[70] Not only its nature, but also its implications and applications are multidisciplinary and diversified. In the recent past, industrial orientation of biotechnology has been phenomenal. Given the fact that the results of such orientation have been promising, the industrial application of biotechnology is expanding day by day. The research and development in the field has proven the potential of biotechnology, resulting in the

---

[67]   See Sreenivasulu. 2005. 'Biotechnology and Patent Law'.

[68]   'WIPO Study on the Protection of Inventions in the Field of Biotechnology'. 1987. EIPO/Cornell University, New York.

[69]   'Study on the Importance of Biotechnology for National Growth and Development'. 1997. Gene Campaign, New Delhi.

[70]   Smith. 1996. *Biotechnology*, p. 6.

investment of huge amounts of money in the field. In fact, biotechnology has created formidable belief in the minds of people and driven industries to utilize the potential of research and development. Much of the modern biotechnology has been developed and utilized by large companies and corporations. However, many small and medium-sized companies are realizing that biotechnology is not a science of the future but provides real benefits to their industry today. Of course, biotechnology-based industries are not labor-oriented; these would need minds more than muscles. Industrial application or commercial orientation of biotechnology has been seen because of academia–industry collaboration. Not only industry, but government agencies also have collaborated with academics for research ventures. In the developed nations, in particular, in the US, the National Institute of Health (NIH) funds genetic engineering and DNA-related research[71] through academic collaborations with universities and research institutions.[72] Entrepreneurs from academia, who are often dominant charismatic individuals with the primary aim of developing a new technology, have developed many new biotechnology companies. The expansion of biotechnology and its application is necessarily a result of transfer of laboratory research to industrial output where the technology has been put to large-scale industrial application and commercial utilization. The promise and prosperity of biotechnology, and its extensive implications and applications, have given rise to several concerns and issues. These include the promotion of research and development in the field, protection of intellectual property rights, regulation of use and development of biotechnology, monitoring the safety in the use of technology and its products, supervision of trade in biotechnology and its products, policy considerations in biotechnology, biological and ecological concerns involved in biotechnology, public policy issues, societal and religious concerns, and many others. With the given background, the subsequent chapters address potential issues and concerns, and debate on the evolving policy framework in this regard toward establishing laws related to biotechnology.

---

[71] See generally, Rockman, H.B. 2004. *Intellectual Property Law for Engineers and Scientists*, p. 259. NJ, USA: Wiley–Interscience.

[72] Kornberg. 1995. *The Golden Helix: Inside Biotech Ventures*, p. 11.

3 Evolving a Policy Framework for Biotechnology

Ingenuity shall receive liberal encouragement. The state should promote the progress in science and technology for the betterment of the lives of people and the community at large. The language of the basic norms and constitutions of the nations expresses concerns over the importance of the progress of science and technology for the society. The United States Constitution states that the Congress shall promote science, technology, and arts with an obligation to encourage research and development in these fields by granting temporary monopoly to the inventors and authors of intellectual work. In the US, the state undertakes the responsibility of stimulating the progress of science and technology by the conferment of exclusive monopoly to the inventors and authors.[1] The state also gets involved in the research by funding and undertaking

---

[1] See Sreenivasulu, N.S. 2005. 'Biotechnology and Patent Law', *Journal on World Intellectual Property Rights*, 1(1–2, July–December).

the research and development in different sectors. However, it cannot cater to the needs of the society on its own and hence assigns its obligations to different sectors including the university, academia, and the private researcher. In such a situation, it becomes inevitable to provide some incentives to the inventor or the author in the form of some reward, compensation, or exclusive monopoly. For overall development, the society depends upon the outcome of research in different fields of arts, science, and technology.[2] Having recognized this fact, few states such as the US undertake funding of research and development in different sectors to provide the necessary stimulation. Enormous potential of research is actually lagging behind due to lack of funds. In fact, in most of the developing countries the status of research and development cuts a sorry figure because of financial constraints. Meanwhile, we find that in developed nations, where there is free flow of finance and revenue, the state either undertakes research on its own or funds research and development as a matter of state policy. These nations also encourage private participation in research and development. The states must promote the research and development in science and technology for the progress of the society. There should be efforts for capacity building in terms of analyzing policy implications of trends and corrective measures required in the field of biotechnology at the global level.[3]

Issues and Challenges

Any policy framework today has to contribute to sustainable development,[4] that is, development that meets the needs of the present generations without compromising the ability of future generations to meet their own needs.[5] The issue of how scientific and technological

[2] See Sreenivasulu, N.S. 2013. 'Law and Policy of Science and Technology in India,' a policy document and report submitted to MHRD, Government of India through the funding of Distance Education Council and Karnataka State Open University, Mysore.

[3] Chaturvedi, S. and S.R. Rao. 2004. *Biotechnology and Development: Challenges and Opportunities for Asia*, p. 29. New Delhi: Academic Foundation.

[4] Wuger, D. and T. Cottier (eds). 2008. *Genetic Engineerig and the World Trade System: World Trade Forum*, p. 5. UK: Cambridge University Press.

[5] Report of the World Commission on Environment and Development, 'Our Common Future' (Brundtland Report), UN Doc. A/42/427(1987).

progress impact the law is not new. In the past, we have seen scientific and technological advancements posing challenges to the cannons and cantors of law, and we had successfully adopted, adjusted, or restructured the law to address the complexities raised by such challenges. In this regard, biotechnology is no different and many of its issues and challenges need to be addressed in the realm of law and policy. There are some vital policy options and key areas of biotechnology that would require governance under law, such as: providing market access to the biotechnology products; ensuring international strategic biotechnology alliances and cooperation; and protection and risk management in terms of regulating the exposure of biotechnology products to the human and the environment.[6] In the field of biotechnology, however, the process of adaptation of the law to scientific and technological advances seems particularly problematic. There are a number of issues to be addressed in formulating the biotechnology regulatory policy, which are actually quite challenging to deal with. First, although there is general agreement over the need for regulation in the different fields of biotechnology application such as agriculture, medicine, food production, animal husbandry, beverages, energy, equipment, environment protection, chemical sciences, forensic science, life sciences, and such other fields, domestic laws have been deeply divided over the approach to be taken. The disagreement is on whether regulation should be of the "command and control" type or be based on flexible codes of conduct or be left to self-regulation by relevant stakeholders. Second, ethical standards play a fundamental role in determining the limits of the permissible scope of scientific research and directing scientific progress toward socially and morally acceptable goals, but the diversity of such standards hinders a common understanding of such limits and goals.[7] Third, biotechnology inventions might cause a problem for the safety of the environment and the ecological balance and diversity if not handled and used properly, but the approach to regulate the improper use and release of biotechnology products into the environment is not uniform. Fourth, biotechnologically produced food (genetically modified food)

[6] See Ricardo Melendez, O. and V. Sanchez. 2007. *Trading in Genes: Development Perspectives on Biotechnology, Trade and Sustainability*, p. 11. London: Earth Scan.

[7] See Francioni, F. and T. Scovassi (eds). 2006. *Biotechnology and International Law*, p. V. Oxford and Portland, Oregon: Hart Publishing.

with nonnatural qualities and traits needs to undergo proper tests and trials before it is released for consumer use. The approach on labeling requirements in such cases regarding the contents, ingredients, and the results of tests and trials needs to be uniformly accepted. Fifth, biotechnology research, innovations, and products might go against certain religious sentiments; in such cases, on what basis shall the parity between the religious sentiments with research be developed is a question that needs to be answered. Sixth, biotechnology activities are not carried out by the traditional subjects of international law, the states, but rather by private actors in science and business who are not readily amenable within the scope of application of the traditional categories of international norms that are designed to regulate state action. Seventh, the issue of whether biotechnology research leading to novel innovations and creations, which include living beings with genetic modifications, needs to be promoted and protected through private monopoly needs to be addressed. How and how far such innovations shall be protected under intellectual property regime and whether such protection goes against public interest are other fundamental questions. Protection and promotion of biotechnology involve addressing these pertinent issues and challenges. The amount of development in biotechnology revolution we have seen in the past three decades seems quite promising in terms of appreciating the research and development in biotechnology as well as addressing the challenges arisen therefrom. The aim is to promote biotechnology while ensuring its safe and proper use for sustainable development. Nations have attempted to appreciate the same in their respective science and technology policies and strategies. Science and technology policies at the domestic level would play a vital role in promoting and protecting the biotechnological innovation. Of course, the approaches for harmonizing socioeconomic requirements and agenda for science and technology policy would have to be synergized for optimum utilization of resources.[8]

Promotion of Science and Technology as a State Policy

Lack of political commitment combined with limited financial and human resources, lack of consumer and producer support, low capacity,

---

[8] Chaturvedi and Rao. 2004. *Biotechnology and Development*, p. 29.

and high entry costs have been posing serious problems for policy initiatives for promoting science and technology, including biotechnology.[9] Having recognized the potential of science and technology in the progress of the economy and overall development of the nations, the states do formulate special policies to fund and undertake research and innovation, at least in the areas of utmost importance and public interest.[10] The growth of most of the developed nations is driven by the development of technology and progress of science, and these nations now formulate policies to sustain and regulate the growth and development in science and technology. However, the developing nations, which are struggling with the lack of growth in the field of science and technology, are trying their best to stimulate the research and development by providing incentives for research and innovations. There is a marked difference between the policies of developing and developed nations in terms of fund allocation, infrastructural facilities, and knowledge pool creation and usage. Therefore, the structure and sustenance of the policy vary depending on the status of the country in terms of economy, and scientific and technological advancements. While the Government of India[11] has adopted general science and technology policies, there are also biotechnology-specific policies in addition to them. The general policies aim to promote and develop every area of science and technology such as industrial technology, information technology, medical science, physical sciences including nuclear technology, space technology, chemical sciences including biochemical sciences, material sciences, earth sciences including geology, geography, and oceanography, life sciences including botany, zoology, biology, microbiology, biotechnology, genetics, home sciences including nutrition, and mathematical sciences including statistics. However, noticing greater significance, influence, and utilities of biotechnology and information technology for society, specific policies are adopted to focus exclusively on their promotion and growth.

[9] See Ricardo Melendez and Sanchez. 2007. *Trading in Genes*, p. 37.

[10] See Sreenivasulu. 2013. 'Law and Policy of Science and Technology in India'.

[11] Government of India under the aegis of Ministry of Science and Technology adopted Biotechnology Strategy and Policy in 2003.

Constitutional Background of Promotion of Science and Technology

From a constitutional perspective, legitimate national processes and policies of decision-making on science and technology should not be circumvented by international regulatory activities.[12] These issues should be decided by the national legislative through the respective constitutional frameworks without or with less international interferences. In general, the states recognize the importance of promotion of science and technology, including biotechnology, by envisaging the same in their respective constitutions through proper legislations. For instance, the US Constitution stresses on the importance of promotion of science and technology for the prosperity of the society,[13] thus empowering the US government to take any measure for promotion and sustenance of research in science and technology. Furthermore, different departments of the federal government of the US as well as the states undertake separate policies for the promotion of science and technology in their respective governing regions. India also envisaged the promotion of science and technology as a policy of the state in its constitution. Although the Constitution of India, unlike the US Constitution, does not specifically empower the Government of India to take any measure to promote science and technology though the careful reading and understanding of the different provisions of the Indian constitution show that it emphasizes on the promotion and development of science and technology. For instance, the fundamental rights and duties of the people of India listed under Part III and Part IV of the Constitution of India imply how Indian government is promoting science and technology among its citizens.[14]

Fundamental Duties

The Fundamental Duties under Article 51A of the Constitution of India obligate citizens to develop scientific temper. In 1976, through

---

[12] See Wuger and Cottier (eds). 2008. *Genetic Engineering and the World Trade System*, p. 16.

[13] See Sreenivasulu. 2005. 'Biotechnology and Patent Law'.

[14] See Sreenivasulu, N.S. 2007. *Human Rights: Many Sides to a Coin*, 1st edn. New Delhi: Regal Publications.

42nd amendment to the Constitution of India, a new chapter on Fundamental Duties was added. It was felt that the Constitution of India listed various fundamental rights but did not say anything about fundamental duties. However, the philosophy of rights is that they are always coupled with duties. Recognizing the need to have constitutionally recognized duties, the Government of India amended the Constitution of India. After this amendment, Part III speaks about fundamental rights, Part IV about directive principles of state policy, and Part IVA about fundamental duties. Part IVA of the constitution, added later as part of the amendment, consists of Article 51A, which specifies a code of 10 fundamental duties of the citizens.[15] One of the fundamental duties of the citizens of India is "development of scientific temper" and to strive toward excellence in all spheres of individual and collective activity so that the nation constantly rises to higher levels of endeavor and achievements.[16] It does not mean that every citizen should become scientist or researcher working in a laboratory, but it does mean that there shall be possible contribution of every Indian to the progress of science and technology and, in turn, the progress of the Indian society.[17] Therefore, it can be said that the Constitution of India envisages the development of science and technology by prescribing this particular duty for its citizens. It intends to create an environment that is favorable to the promotion and development of science and technology through the support of the citizens.

Fundamental Rights

Furthermore, the Constitution of India provides freedom of speech and expression to its citizens.[18] Since scientific and technological developments arise from innovative thoughts and expression of the same, freedom of speech and expression has been read to encompass freedom

---

[15] See Pandey, J.N. 2007. *Constitution of India*, 44th edn. Allahabad: Central Law Agency.

[16] See Constitution of India, Article: 51A (j).

[17] See generally, Singh, M.P. and V.N. Shukla. 2013. *Constitution of India*, 12th edn. Allahabad: Eastern Book Company.

[18] Sreenivasulu, N.S. and Somashekarappa. 2012. 'Freedom of Speech and Expression, Intellectual Property and Copyright', *Manupatra Intellectual Property Reports*, II(I, June).

of thought, expression, and execution.[19] Research and innovation involve conceiving of ideas and execution of the same resulting in technological and scientific advancements. Individuals are encouraged to enjoy freedom of speech and expression, which could result in innovations and novel creative works in the fields of science and technology. An innovation results out of the expression of an idea by following a scientific method. Therefore, such expressions resulting in useful inventions need to be protected in order to encourage further research and expression of scientific ideas. In fact, different forms of intellectual property rights including patents, copyrights, trademarks, industrial designs, integrated circuits, plant and animal varieties, and trade secrets are nothing but execution of well-conceived ideas in different fields of science and arts. As a matter of fact, fundamental freedom of expression guarantees right to innovate, and freedom to create and seek protection of the same through intellectual property rights. Scientific innovations originate from the expression of ideas that have great commercial and technological value. Although there is no specific or expressive mentioning of the same in the Constitution of India either under fundamental rights and freedom or under any other part, it is assumed that intellectual property rights protection could be traced to and has a connection with the fundamental freedom of speech and expression, which is guaranteed to every citizen of India under Article 19(a).[20] Even the Preamble of the constitution echoes similar terms when it states that the constitution guarantees to its citizen liberty of thought and expression.[21] Liberty of thought follows

[19] Sreenivasulu, N.S. and Somashekarappa. 2013. 'Freedom of Speech and Expression in the Indian Democracy', *International Journal of Law and Policy Review*, 2(2): 219–35.

[20] Sreenivasulu, N.S. 2012. 'An Interface between Intellectual Property Rights and Human Rights', *International Journal of Legal Research Studies*, I (September): 1–12.

[21] The Preamble of the Constitution of India declares:

WE, THE PEOPLE OF INDIA, having solemnly resolved to constitute India into SOVEREIGN SOCIALIST SECULAR DEMOCRATIC REPUBLIC and to secure to all its citizens:

JUSTICE, social, economic and political;
LIBERTY of thought, expression, belief, faith and worship;

conceiving of idea and expression of the same when it comes to scientific innovations. Therefore, it can be said that Article 19(a) read with the preamble of the constitution implies protection of intellectual property rights arising out of the exercise of liberty of thought and conceiving of scientific ideas resulting in innovations once such thoughts and ideas are properly executed through scientific methods.

The Ministry of Science and Technology

India, as a developing nation (as an underdeveloped nation at the time of the framing of the constitution), could not provide specific policies and funding for science and technology compared to the developed nations. However, the Government of India has a separate ministry under the central government,[22] namely, the Ministry of Science and Technology, which undertakes the promotion and regulation of science and technology in different spheres. The ministry formulates different policies for the promotion of science and technology, with the latest being the Science and Technology Policy of India, 2013.[23] With the inception of a separate ministry proper amenities and incentives are expected for the promotion, regulation, and sustenance of science and technology. In this connection, the words of Kapil Sibal, the then Minister for Science and Technology, Government of India, are worth quoting:

> Since independence, the Government of India has been strongly aware of both needs—the need to build up a powerful science base, and the need to ensure that science is not restricted to the university laboratories. Under a succession of enlightened leaders, Indian governments have long recognized the need for any country that aspires to call itself a modern nation

---

EQUALITY of status and opportunity;
and to promote among them all
FRATERNITY assuring the dignity of the individual and the unity and integrity of the Nation.

[22] States too have similar state ministries for science and technology apart from the central ministry, which is an encouraging trend.

[23] Department of Science and Technology. 2013. 'Science and Technology Policy', Ministry of Science and Technology, available at http://www.dst.gov.in/st-system-india/science-and-technology-policy-2013 (last accessed 28 February 2016).

to invest heavily in science and technology. The fruits of this foresight are now widely visible. Thanks largely to the government's determination that the country should build a strong independent base in science and technology, India has been able to build up a capacity in a wide range of areas of modern technology, from software engineering to health biotechnology. And this has placed it in a strong position to engage in the global knowledge economy, rather than remaining on the margins.[24]

The institutions receiving grants from the Ministry of Science and Technology are encouraged to retain benefits arising out of research undertaken while utilizing the given grants.[25] Intellectual property and patent rights stemming from the successful research and development endeavors through ministry funding may be kept by the institutions. In case of joint research by institutions, such rights and benefits may be held and enjoyed jointly. The intention has been to see that research and development takes place at the ministry-funded institutions, which generates good intellectual property rights and accordingly commercial benefits. It is also to encourage technology transfer, trade, and commercialization of research and development for the larger benefits of the society to cater to various needs including food and health.

The US Scenario

In countries such as the US, the constitution expressively undertakes the promotion of science, arts, and technology. We find specifications with regard to science and technology and their promotion in the letter of the constitution, which empowers the state to undertake measures to implement the mandate of the constitution. Article I, Section 8 of the constitution of the US states that[26] the congress shall have power to promote the Progress of Science and useful Arts, by securing monopoly for limited time to authors and inventors to their respective writings and

[24] Speech by Kapil Sibal, Science and Technology Policy of India, dated 3 October 2013, available at http://missionras.blogspot.in/2013/10/science-and-technology-policy-of-india.html (last accessed 20 June 2016).

[25] Chaturvedi and Rao. 2004. *Biotechnology and Development*, p. 134.

[26] Article I, Section 8 of the Constitution of America enumerates the different powers of the Congress.

discoveries. Although the US constitution is short and specific in terms of its content, language, and the number of articles and sections compared to the Indian constitution, it has laid down certain principles with regard to undertaking promotion of science and technology. Besides, the US constitution is one of the oldest constitutions in the world today[27] that has been able to visualize the growth of the society along with the growth of science and technology. The US constitution came into being back in the eighteenth century when India was under the British Empire. The Indian constitution, therefore, is very young compared to the US constitution. Understandably, the growth of science, arts, and technology in the developed nations is more advanced, and similarly, the vision of the constitution, legislations, and the mechanism of the state would also be in tandem. In a developing nation that is finding it difficult to manage the economy and provide minimum basic facilities to its citizens, the expectation to have similar policy to promote science and technology would be a bit too much. Being a developed nation, the US not only is able to see the growth of science and useful arts, but also has vision and mission for its promotion and policy implementation for useful and commercial results of any technology, including biotechnology.

State's Obligation to Promote Science and Technology

The state has got the power coupled with an obligation to take all necessary actions and measures to promote and sustain the growth of science and technology and see that it benefits the people and the society. However, in India, the state is under no direct or express obligation to promote and encourage research and development in any field. The promotion of science and technology could be read into some provisions of the constitution through judicial interpretations or pronouncements. The strong intellectual-property-related regulations of the US stem from the power of the congress to promote science and arts conferred upon it by the constitution. The judiciary of the US also, many a times,

---

[27] Although the US constitution is comparatively older, it has undergone only 27 odd amendments till date compared to the Indian constitution, which has been amended as many times as more than 100 times till date since its coming into being in 1950.

reiterated that intellectual property laws derive validity and legality from the Article 1, Section 8 of the constitution.[28] By virtue of this power, the congress has enacted several legislations to create legal framework to provide monopoly for a limited period for research and innovation. One of the philosophies of the US constitution in this regard and the intellectual property legislations in the US that "ingenuity shall receive liberal encouragement" scores highly in terms of promotion of science and technology. Law courts in the US have always advocated for the propagation of this philosophy to promote research and innovation. Even Thomas Jefferson, the famous agriculturalist, scientist, advocate, and senator who drafted first-ever patent law in the US, strongly advocated for the liberal encouragement of ingenuity and intellect in research, leading to creative innovations.[29] Many a times, in cases of ambiguity in the intellectual property legislations or in case of confusion on the intent of the congress with regard to a particular issue, the actual words of the constitution have been taken as an aid. In particular, while dealing with unforeseen technologies such as information technology, biotechnology, nanotechnology, biomedical technology, and space technology, the judiciary and the executive have taken the spirit of the constitution in interpreting the respective legislations to deal with unforeseen technologies and their eligibility for protection under intellectual property regime.[30] The power of the congress to promote science and useful arts has been discussed and debated many times before the judiciary and the intellectual property offices while dealing with novel technologies such as biotechnology. The constitution does not say anything directly

[28] There are several intellectual property legislations congress has enacted by virtue of powers conferred under Article 1, Section 8. For instance, the Patents Act, the Copyrights Act, the Trademarks Act, the Digital Millennium Copyright Act, Plant Patent Act, Plant Varieties Act, and so forth, have been enacted by the Congress to promote and regulate the research and innovations in science, useful arts, and technology.

[29] See Sreenivasulu. 2005.'Biotechnology and Patent Law'.

[30] See *Parker v. Flook* 437 U.S. 584 (1978), *Diamond v. Chakaraborty* 447 U.S. 303 (1980), *Diamond v. Dier* 450 U.S. 175 (1981) where constitutional mandate was taken as aid to interpret the provisions of the US Patent Act to decide the patentability of unforeseen technologies such as biotechnology and information technology.

regarding the eligibility of innovations of biotechnology for protection and promotion. Besides, the respective intellectual property legislations including patent law and copyright law were initially not advocating for the protection of unforeseen technologies including biotechnology. However, it is judicial innovative interpretation of the provisions of the respective legislations that earned eligibility to the unforeseen technologies such as biotechnology for patent, copyright, and other types of intellectual property protection. The judiciary of the US is of the opinion that science and technology needs to be promoted and ingenuity shall be liberally encouraged. In fact, in many of its important decisions that have marked the progress of not only science and technology but also the protective legal mechanism for conferring intellectual property rights, the Supreme Court of US has directed the congress, the legislature, the intellectual property offices, and the executive to follow the principles and philosophies of constitution and the intellectual property legislations where the state is obligated as well as empowered to promote science and technology by means of conferring exclusive monopoly such as intellectual property rights to the inventors and authors for a temporary period of time. The Supreme Court of US, in its landmark decision of *Diamond* v. *Ananda Chakraborty*[31] had said that "anything under the sun made by man" is patentable.[32] The outcome was that if the research and innovation produced a thing that could be anything that marked the human intellect or ingenuity, it should be protected and promoted through the grant of patent or other intellectual property rights. The patent and trademark office of the US also promotes the same philosophy when it welcomes the liberal and innovative decisions and interpretations of the judiciary. In fact, the US judiciary has innovative approaches to promoting research and innovation. It involves one innovation promoting the other innovation, on supportive, if not similar, lines. The congress, judiciary, as well as the executive have been using innovative approaches to promote science and technology in the making,

[31] 1980 US, SC 404.

[32] *Diamond* v. *Anandha Chakraborty* has been the spirit behind the development in research and scientific technologies such as biotechnology. The decision also helped in formulating legal framework for protection and regulation of biotechnology. It may not be untrue if it is said that the case is the foundation for the whole of the biotechnology law.

interpretation, and execution of law. Protection of innovation and intel‐
lectual property has been the primary focus of the US trade policy.[33]

Science and Technology Policy of India

The Government of India has been very active in the formulation of the
science and technology policy, which intends to promote innovation on
one hand and find solutions to major problems in the country, such as
hunger, health, and poverty, on the other. Over 100 global companies
have come to India to set up research and development (R&D) centers,
confirming the intellectual capital of Indian scientific and engineering
community. Science must grapple with the key challenges the country
is facing today. These include the pressures of increasing population,
greater health risks, changing demographics, degraded natural resources,
and dwindling farmlands. We need new science and technologies, new
priorities, and new paradigms to address these fundamental challenges.
We, in India, are practicing new physics and chemistry approaches to
make new materials, which are of direct relevance to the Millennium
Development Goals of the United Nations. The former President of
India, A.P.J. Abdul Kalam, said:

> [T]oday India has become one of the strongest in the world in terms of
> scientific manpower in capability and maturity. Hence, we are in a posi‐
> tion not only to understand the technologies that we may have to borrow,
> but also to create our own technologies with extensive scientific inputs
> of indigenous origin. Basically we have come a long way since our inde‐
> pendence, from mere buyers of technology to those of who have made
> science and technology as an important contributor for national develop‐
> ment and societal transformation. The role of the government and the
> public sector is indeed of real significance in the process of promotion of
> science and technology for national development.[34]

---

[33] Lehman B.A. 2007. 'Making the World Safe for Biotech Patents', *Journal
of Bio-Law and Business*, 6(1): 52–7.

[34] Abdul Kalam, A.P.J. 2005. 'Innovativeness and Foresight of Science Today
to Compete the Developed World of the Future', Ministry of Communication
and Information Technology, available at http://pib.nic.in/newsite/erelcontent.
aspx?relid=8276 (last accessed 20 June 2016).

The government has been promoting the science and technology through formulating policies.[35] In a world where the powers are determined by their share of the world's knowledge, reflected by patents, papers, trade volumes, and so on, the World Trade Organization (WTO) plays a crucial role in the economic development. It is important for India to get its acts together to become a continuous innovator and creator of "science-and-technology-intensive products." India has been responsible for certain scientific breakthroughs since ancient times.[36] India's contribution to the science and technology is notable. However, India has also been known for its brain drain due to lack of support to the research and development. Notably after Independence, the Government of India started showing concern for the development of science and technology. Developing nations, including India, aspire for technology transfer from the developed nations for their various needs, the reasons being underdeveloped sphere of science and technology, high cost of production of technology at home, and inability to provide funding for research and development. The government has recognized that the development of indigenous technology would potentially strengthen the country and make it independent in terms of possession and use of technology. In the half century since Independence, India has been committed to the task of promoting science. The key role of technology as an important element of national development is also well recognized. The Scientific Policy Resolution of 1958, the Technology Policy Statement of 1983, the Technology Policy of 2003, and the Science and Technology Policy of 2013[37] have enunciated the principles on which the growth of science and technology in India has been based over the

[35] Visit the official website of the Ministry of Science and Technology, Government of India for the original texts on science and technology policies that the government of India formulated. See www.dst.gov.in (last accessed 29 January 2016).

[36] See Sreenivasulu, N.S. 2013. 'Law and Policy of Science and Technology in India', a policy document and report submitted to the MHRD, Government of India through the funding of Distance Education Council and Karnataka State Open University, Mysore.

[37] Visit the official website of the Ministry of Science and Technology, Government of India for the latest updates in this regard. See www.dst.gov.in (last accessed 29 January 2016).

past several decades.[38] These policies have emphasized self-reliance, as also sustainable and equitable development. They embody a vision and strategy that are applicable in today's scenario, and would continue to inspire us in our endeavors. Such encouragement and support has led to a sound infrastructural base for science and technology, including research laboratories, higher educational institutions, and highly skilled human resources. Indian capabilities in science and technology cover an impressive range of diverse disciplines, areas of competence, and applications. India's strength in basic research is recognized internationally.

India's success in the fields of agriculture, healthcare, chemicals and pharmaceuticals, nuclear energy, astronomy and astrophysics, space technology and applications, defense research, biotechnology,[39] electronics, information technology, and oceanography are widely acknowledged. Major national achievements include a significant increase in food production, eradication or control of several diseases, and increased life expectancy of its citizens. While these developments have been highly satisfying, one is also aware of the dramatic changes that have taken place, and continue to do so, in the practice of science, technology development, and their relationships with, and impact on, the society. The rapidity with which science and technology is moving ahead is absolutely striking. Science is becoming increasingly inter- and multidisciplinary, and calls for multiinstitutional and, in several cases, multicountry participation. Major experimental facilities, even in several areas of basic research, require large material, human, and intellectual resources. Science and technology have become so closely intertwined, and reinforce each other, such that, to be effective, any policy needs to view them together. The continuing revolutions in the field of information and communication technology have had profound impact on the manner and speed at which scientific information becomes available and scientific interactions take place. Science and technology have had an

[38] See also the official website of the Ministry of Science and Technology, Government of India at www.dst.gov.in (last accessed 29 January 2016).

[39] There is a separate department for biotechnology under the Ministry of Science and Technology, which looks after promotion and protection of biotechnology exclusively: See generally, Chaturvedi and Rao. 2004. *Biotechnology and Development*, p. 53.

unprecedented impact on the economic growth and social development. Knowledge has become a source of economic might and power. This has led to increased restrictions on sharing of knowledge, new norms of intellectual property rights, and global trade and technology control regimes. Scientific and technological developments today also have deep ethical, legal, and social implications. Globalization and the intensely competitive environment have a significant impact on the production and services sectors. As a result, our science and technology system has to be infused with a new vitality if it is to play a decisive and beneficial role in advancing the well-being of all the sections of our society. The nation continues to be firm in its resolve to support science and technology in all its facets. It recognizes its central role in raising the quality of life of the people of the country, particularly of the disadvantaged sections of society, in creating knowledge and wealth for all, in making India globally competitive for trade and commerce, in utilizing natural resources in a sustainable manner, in protecting the environment, and for ensuring national security.[40]

Objectives Set Forth in Science and Technology Policy

The Constitution of India under Article 51A postulates for the development of scientific temper among the citizens of India.[41] It is a fundamental duty of the citizens of India to have a positive and healthy attitude toward science and technology, which shall create an atmosphere for its encouragement. The intention is to integrate universities and the industry for sustained growth in different fields of science and technology to address the various needs of the country in terms of guaranteeing the right to food, right to health, right to work, and the like. The science and technology policy aims to develop indigenous knowledge in various fields to help India become a hub of knowledge, information, and technology. Technology for the human welfare, technology for the maximum benefit of the maximum number of people is the underlined principle

[40] Preamble to the Science and Technology Policy of India, 2003, available at the website of the Ministry of Science and Technology, Government of India at www.dst.gov.in (last accessed 29 January 2016).

[41] See generally, Singh and Shukla. 2013. *Constitution of India.*

of the policy. The basic objectives of the policy[42] could be inferred as follows:

1. Making people aware of the advantages of advanced scientific temper to make India a scientifically progressive and enlightened society.

2. Integration of science and technology with different spheres of national activities and encouraging people to participate.

3. Application of technology in various fields to cater to different needs of the society; guarantee sustainable development in various sectors such as agriculture and medicine, and ensure the safety and security of the environment.

4. Use of technology toward sustainable efforts in eradicating poverty by providing employment and livelihood; to remove regional imbalances; to exploit traditional knowledge pools.

5. Development and screening of technologies for widespread dissemination and networking to reach unorganized sectors.[43]

6. Promoting national strategic and security-related objectives with the latest advances in science and technology.[44]

7. Fostering the research and development in science and technology by driving young people to have their careers in research. Building centers of excellence in various fields of science and technology and raising the standards of research to meet international standards.

8. Necessary freedom and autonomy to academic and research institutions to ease and speed up the progress of science and technology, and to encourage creative work with social responsibility.

---

[42] The latest science and technology policy was adopted in January 2013 by the Government of India, which is preceded by science and technology policies of 2003, 1983, and 1958.

[43] For further details, please visit the official website of the Ministry of Science and Technology, Government of India at www.dst.gov.in (last accessed 29 January 2016).

[44] See Sreenivasulu, N.S. 2013. 'Law and Policy of Science and Technology in India', a policy document and report submitted to the MHRD, Government of India through the funding of Distance Education Council and Karnataka State Open University, Mysore.

9. Strengthening research bases and mechanism related to the evaluation, absorption, and upgradation of technology.

10. Use of science and technology to utilize, promote, preserve, and advance extensive traditional knowledge.

11. Public and private participation and cooperation in research and innovation, for the promotion of the economy and the society, while giving importance to the key technologies such as biotechnology and information technology.

12. Promotion of research and development in science and technology by providing incentives in the form of intellectual property rights which provide exclusive monopoly to the inventors or authors.

13. Commercialization of innovations in public interest to maximize benefits to the society.

14. Creating quality information and databases useful in the development of science and technology and providing access to information at affordable prices.

15. Development and use of technology to forecaste natural hazards to prevent or mitigate the damage, in particular, floods, cyclones, earthquakes, drought, and landslides.

16. Achieving national development and promoting international cooperation through the progress of science and technology.

17. Integration of science and technology with other disciplines by having insights into the respective disciplines to encourage multidisciplinary integrated approach in research for the application of innovations of science and technology to the prosperity of different disciplines.

18. Involvement of scientists and technologists in national governance and policymaking for having a rightful approach toward promotion and application of science and technology in public interest.

19. Establishing an exclusive department for biotechnology[45] promoting research, development, and commercialization in biotechnology.[46]

---

[45] Visit the official website of Department of Biotechnology at www.dbtindia.nic.in (last accessed 29 January 2016).

[46] Chaturvedi and Rao. 2004. *Biotechnology and Development*, p. 27.

Having set very promising objectives for the promotion of science and technology, the Government of India has a strategic plan to implement these objectives to meet the mission and vision of the policy. It involves people from academia, industries, and policymaking to discuss all the issues and challenges across the board from the ground level to the top level of policy implementation.[47] Efforts have been made to ensure that people who have worked at the ground level of research in science and technology are placed in the respective departments of the government that are responsible in devising policies and plans in order to see that policies and plans could be practically successful. In fact, such a dynamic strategy was required to infuse a new sense of dynamism in the development of science and technology with the participation and contribution of all the stakeholders. There is an advisory body to the Ministry of Science and Technology that consists of scientists and technologists who continuously advice the ministry on its policies and strategies. Perhaps there is also a move to strengthen and equip the respective departments, universities, and research institutions to see the practical and smooth implementation of the policy for prosperity. Compared to research funding by the governments in countries such as the US and the UK, the current funding for research and development in India is too low. There is a need to increase funding to keep the research and development sector free from financial crunches. In light of the situation, the policy has been set out for increased funding for research in science and technology. In the era of multidisciplinary research, the policy aims to fund and promote such researches leading to innovative results. The need for the creation of a comprehensive national system of innovation covering science and technology, as also legal, financial, and other related aspects, has been recognized. There is a need to change the ways in which society and economy perform, if innovations have to fructify. Perhaps, early initiatives of biotechnology transfer of research leads, protocols, and technologies to industry have been spearheaded by the Ministry of Science and Technology under its direct supervision.[48]

---

[47] For further details, please visit the official website of the Ministry of Science and Technology, Government of India at www.dst.gov.in (last accessed 29 January 2016).

[48] Chaturvedi and Rao. 2004. *Biotechnology and Development*, p. 133.

Efforts are being made to drive for industry–academia collaboration, thereby promoting industry participation in academia and vice versa; direct involvement and funding of academic research by industry; and transfer of research output and technology from academia and research to industry. Development and sustenance of traditional knowledge and the use of the same in the scientific and technological research and innovation has been strongly advocated. Since India is a hub of traditional and indigenous knowledge and technologies, India is required to recognize the need to promote the same and make all efforts to encourage further research to develop traditional knowledge. Further development of technologies in pursuit of mitigation and management of natural hazards has been initiated.[49] In the era of globalization where there is more emphasis on private property and private domain than public property and public domain, similar trend can be seen in research and development. There is growing need for strong intellectual property regime to encourage research and innovation with incentives for exclusive and private proprietary rights. The present global order demands for more and more incentives and stimulations that confer private and exclusive monopoly on research innovations. Similarly, following the world order, India also advocates for effective and strong intellectual property regime. In the wake of vanishing of boundaries and the entire world becoming a global village, no state can remain isolated without international collaboration in every field. Recognizing the same, the states have been busy in executing bilateral and multilateral treaties and compacts to have collaborations to boost greater development. The Indian science and technology policy[50] also encourages international collaborations for effective and increased growth and cooperation in science and technology, including mutual cooperation, coordination, and exchange of ideas, equipment, manpower, techniques, and technologies. Joint ventures

[49] See also Sreenivasulu N.S. 2013. Law and Policy of Science and Technology in India, A policy document and report submitted to the MHRD, Government of India through the funding of Distance Education Council and Karnataka State Open University, Mysore.

[50] For further details, please visit the official website of the Ministry of Science and Technology, Government of India at http://dst.gov.in/ (last accessed 29 January 2016).

in research and innovation are encouraged with joint action for trade and commercial benefits. States are not only stressing for international cooperation but also keen on developing local and domestic cooperation in terms of participation and contribution from different corners and quarters as well as sectors of industry, departments, academia, organizations, people, and local governments. Besides, the states have undertaken sensitization orientation programs for stakeholders as well as people in general to develop a sense of understanding among the masses on the potential use of science and technology and also to evolve cooperation and coordination from the general public at large. With emphasis on keeping research and development free from financial difficulties, there is increased flow of funds. Direct funding and financing of applied or advanced research by industry and corporate sectors, as well as private and public funding, is encouraged. All in all the vision and mission of Indian science and technology policy aims to build a new and resurgent India that continues to maintain its strong democratic and spiritual traditions, and remains secure militarily, socially, scientifically, technologically, and economically. It has been recognized that it is important to draw on the many unique qualities of the Indian civilization that define the inner strength of India. This has been intrinsically based on an integrated and holistic view of nature and life with an intention to execute Indian policy to be in harmony with our world view of the larger human family all around. It will ensure that science and technology truly uplifts the Indian people and indeed all of humanity.

Exclusive Biotechnology Policy: Issues and Challenges

Investments in biotechnology in India have begun with the early initiatives of the government and setting up of a separate Department of Biotechnology.[51] Biotechnology, globally recognized as a rapidly emerging and far-reaching technology, is aptly described as the "technology of hope" for its promise of food, health, and environmental sustainability. The recent and continuing advances in life sciences clearly unfold a scenario energized and driven by the new tools of biotechnology. From a constitutional perspective, legitimate national processes and policies

---

[51] Chaturvedi and Rao. 2004. *Biotechnology and Development*, p. 135.

of decision-making on biotechnology should not be circumvented by international regulatory activities.[52] Since there are no uniform acceptance levels for biotechnology across the globe, the issues pertinent to policy regulation of biotechnology should be decided by the national legislative through the respective constitutional frameworks. The rapid development of biotechnology in the last 30 years, with applications to human health and reproduction, agriculture, insurance, and security sectors, has generated varied policy responses from governments in the Organisation for Economic Co-operation and Development (OECD) countries. Though often labeled biotechnology regulation, the vast bulk of the policy literature is concerned with the construction of only one element of a regulatory regime—the normative structure of principles, standards, and rules. Biotechnology regulation, as a field of public policy, has not yet matured to the point where other elements of regulatory regimes, notably processes for monitoring and mechanisms of behavioral modification, are routinely considered or problematized.[53] Thus, the Government of India has initiated certain measures to promote biotechnology. Perhaps, the year 1982 could be mentioned as a landmark year in the history of biotechnology in India. In order to promote biotechnology, an agency called the National Biotechnology Board (NBTB) was constituted under the Ministry of Science and Technology. It is an apex coordinating body to identify priorities, plan, coordinate, and oversee research and development, human resource development, and industrial development of biotechnology. The same board became Department of Biotechnology at a later stage.[54]

Why Biotechnology is Important to India

The Indian biotechnology sector is gaining global visibility and is being tracked for emerging investment opportunities. With encouragement from the government, the Indian biotechnology sector is progressing

---

[52] See Wuger and Cottier (eds). 2008. *Genetic Engineering and the World Trade System*, p. 16.

[53] Somsen, H. 2007. *The Regulatory Challenge of Biotechnology: Human Genetics, Food and Patents*, p. 19. UK: Edward Elgar.

[54] Chaturvedi and Rao. 2004. *Biotechnology and Development*, p. 128.

equally if compared with the developed nations, human capital being perceived as the key driver for global competitiveness. The reason for decline in the development of biotechnology sector in the developed countries is a decreasing appetite for risk capital in the regions where survival lifelines are being provided by the lower-cost research environment of the developing world, such as India. For a country like India, biotechnology is a powerful enabling technology that can revolutionize agriculture, healthcare, industrial processing, and environmental sustainability. The Indian biotechnology sector, over the past two decades, has taken shape through a number of scattered and sporadic academic and industrial initiatives. The time is now ripe to integrate these efforts through a pragmatic National Biotechnology Development Strategy. It is imperative that the principal architects of this sector along with other key stakeholders play a concerted role in formulating such a strategy to ensure that we not only build on the existing platform but also expand the base to create global leadership in biotechnology by unleashing the full potential of all that India has to offer. Biotechnology can deliver the next wave of technological change that can be as radical and even more pervasive than that brought about by the information technology (IT). Employment generation, intellectual wealth creation, expanding entrepreneurial opportunities, and augmenting industrial growth are a few of the compelling factors that warrant a focused approach for this sector. Biotechnology, as a business segment for India, has the potential of generating revenues to the tune of USD 5 billion and creating more than one million jobs by 2020 through products and services.[55]

[55] This can propel India into a significant position in the global biotechnology sweepstakes. Biopharmaceuticals alone have the potential to be a USD 2 billion market opportunity largely driven by vaccines and bio-generics. Clinical development services can generate in excess of USD 1.5 billion, while bio services or outsourced research services can garner a market of USD 1 billion over this time scale. The balance USD 500 million is attributable to agricultural and industrial biotechnology. India has many assets in its strong pool of scientists and engineers, vast institutional network, and cost effective manufacturing. There are over a hundred National Research Laboratories employing thousands of scientists. There are more than 300 college-level educational and training institutes across the country offering degrees and diplomas in biotechnology,

India is reorganized as a mega biodiversity country, and biotechnology offers opportunities to convert our biological resources into economic wealth and employment opportunities. Innovative products and services that draw on renewable resources bring greater efficiency into industrial processes, check environmental degradation, and deliver a more bio-based economy. Indian agriculture faces the formidable challenge of having to produce more farm commodities for our growing human and livestock population from diminishing per capita arable land and water resources. Biotechnology has the potential to overcome this challenge to ensure the livelihood security of 110 million farming families in our country.

Scope and Future of Biotechnology in India[56]

The advancement of biotechnology as a successful industry confronts many challenges related to research and development, creation of investment capital, technology transfer and technology absorption, patentability and intellectual property, trade and commerce, consumer goods, affordability in pricing, safety and regulatory issues, and public confidence. Central to this are two key factors: affordability of and accessibility to the products of biotechnology. India needs to formulate policies that foster a balance between sustaining innovation and facilitating technology. There are several social concerns that need to be addressed in order to promote biotechnology innovation in our country such as conserving bioresources and ensuring safety of products and processes. Government and industry have to play a dual role to advance the benefits of modern biotechnology while at the same time educate and protect the interests of the public. Wide utilization of new technologies would require clear demonstration of the new value addition to all stakeholders. The National Science and Technology Policy of the

---

bioinformatics, and the biological sciences, educating nearly 500,000 students on an annual basis. More than 100 medical colleges add 17,000 medical practitioners per year. About 300,000 postgraduates and 1,500 PhDs qualify in biosciences and engineering each year. These resources need to be effectively marshaled, championed, and synergized to create a productive enterprise.

[56] India is gradually becoming a key player in the biotechnology sector with its vast human resources and huge knowledge pool.

Government and the Vision Statement on Biotechnology issued by the Department of Biotechnology have directed notable interventions in the public and private sectors to foster life sciences and biotechnology.[57] The emergence of India as a global player in the biotechnology sector requires the government to play the role of a facilitator and foster an international competitive environment for investment and enterprise development. India's strategy must be to get more value from its R&D investment, Intellectual Property Rights (IPRs) generation, and tradable goods and commodities. With around 200 industries, the growth of the biotechnology sector in India has been rapid, promising to fulfill India's strategies.[58] India has to develop its own biotechnological and pharmaceutical products to ensure quality and affordability for global trade. In addition to opportunities in drug discovery and development, there are significant openings to provide services to the worldwide biotechnology and pharmaceutical industries and to leverage low-cost high-quality manufacturing with a global discovery potential. Capitalizing on these opportunities would create many new valuable jobs in India as we have seen in the outsourcing and service industry.[59] The vision is to maximize

[57] There has been substantial progress in terms of support for R&D, human resource generation, and infrastructure development over the past decade. With the introduction of the product patent regime, it is imperative to achieve higher levels of innovation in order to be globally competitive.

[58] Current estimates indicate that the industry grew by 39 percent annually to reach a value of USD 705 million in 2003–4. Total investment also increased in 2003–4 by 26 percent to reach USD 137 million. Exports presently account for 56 percent of revenue. Currently the biopharma sector occupies the largest market share of 76 percent followed by bio-agri 8.42 percent, bioservices 7.70 percent, industrial products 5.50 percent, and bioinformatics 2.45 percent. The bioservice sector registered the highest growth (100 percent) in 2003–4 with bioagri 63.64 percent and biopharma 38.55 percent. The policy review has envisaged an annual turnover of USD 5 billion in the year 2010.

[59] However, to achieve the targeted business volume, several new challenges have to be met. These are predictable and enabling policies; increased public and private support for early or proof-of-concept stage of product development; improved communication among stakeholders in the sector; public–private partnerships; integration of the Indian biotechnology sector globally; and improved infrastructure.

opportunities in the area of contract research, manufacturing, and to promote discovery and innovation. The biotechnology industry, being capital-intensive in nature, has historically relied on venture capital from public and private sources. India needs to provide active support through incubator funds, seed funds, and provision of various incentives in order to develop the sector. In a highly competitive and fast-moving business environment, innovative capacity is an important determinant of the ability to create a continuing pipeline of new products and processes. Innovation covers knowledge creation (R&D), knowledge diffusion (education and training), and knowledge application (commercialization). Innovation[60] is not a one-time event; instead, it has to respond continuously to changing circumstances for sustainable growth. The challenge now is to join the global biotechnology league. This will require larger investments and an effective functioning of the innovation pathway. Capturing new opportunities and the potential economic, environmental, health, and social benefits will challenge government policy, public awareness, and educational, scientific, technological, legal and institutional frameworks.[61] The issue of access to the biotechnology products in medicine, agriculture, and industrial sector is of paramount importance. Therefore, there should be adequate support for public-good research designed to reach the unreached in terms of technology empowerment. Both "public good" and "for-profit" research should become mutually reinforcing. Public institutions and industry both have an important role in the process. The National Biotechnology Development Strategy takes stock of what has been accomplished and provides a framework for the future. The policy framework is a result of

---

[60] Innovation is measured in terms of external/domestic patent applications; human capital devoted to R&D; government expenditure on R&D proportionate to country's GDP; business-funded expenditure on R&D; indigenous technologies standardized, demonstrated, and transferred to industry for commercialization; and the number of spin-off companies created. Clear government policies for promotion of innovation and commercialization of knowledge will propel the growth of the biotechnology sector.

[61] Public awareness is increasingly growing with reference to the implications of biotechnology with the active participation of public in the biotechnology-related activities and equally because of proactive media in the contemporary times.

wide consultation and consensus with stakeholders, including scientists, educationists, regulators, representatives of society, and others.[62] This policy aims to chalk out the path of progress in sectors such as agriculture and food biotechnology, industrial biotechnology, therapeutic and medical biotechnology, regenerative and genomic medicine, diagnostic biotechnology, bioengineering, nano-biotechnology, bioinformatics and IT-enabled biotechnology, clinical biotechnology, manufacturing and bioprocessing, research services, bioresources, bioenergy, and environment. The Government of India is concentrating on capacity building in terms of human resource and sophisticated infrastructure for research and development.[63] Several state governments have enunciated biotechnology policies spelling out a comprehensive blueprint for the sector. It is, therefore, prudent to have a National Biotech Development Strategy that charts an integrated 10-year road map with clear directions and destinations. This is the time for investment in frontier technologies such as biotechnology. It is envisaged that clearly thought-out strategies will provide direction and enable action by various stakeholders to achieve the full potential of this exciting field for the social and economic well-being of the nation. In the year 2012, a bioenergy road map was developed, which aimed to lay down the policy framework and facilitation of creating bioenergy and fuel. Since energy is the lifeline of any industrial sector and even for the human habitation, it was thought that biotechnology-enabled bioenergy could cater to the needs and sustainability.

National Biotechnology Policy and Strategy in India, 2014

The need for an integrated biotechnology policy with sufficient attention to different subsets of the sector such as health, agriculture, environment, industry, and other application areas is a prerequisite for the progress of Indian biotechnology sector. The National Science and Technology Policy of the Government and the Vision Statement on Biotechnology

---

[62] It focuses on cross-cutting issues such as human resource development, academic and industry interface, infrastructure development, lab and manufacturing, promotion of industry and trade, biotechnology parks and incubators, regulatory mechanisms, public education, and awareness building.

[63] See Chaturvedi and Rao. 2004. *Biotechnology and Development*, p. 129.

issued by the Department of Biotechnology (DBT)[64] provide a frame-work and give strategic direction for different sectors to accelerate the pace of development of biotechnology in India. The policy has given meaning to the efforts of public and private sectors, the key being a quad-rilateral agreement between universities, industry, laboratories working in the field, and the state. The Biotechnology Policy and Strategy has been a comprehensive attempt of the Government of India in formulat-ing the legal framework for the regulation of promotion and protection of biotechnology in India in various facets.[65] The Indian government adopted the new National Biotechnology Policy and strategy in 2014, which is considered to be an improvement over the earlier policy of 2008. The main features of the policy could be highlighted as follows:[66]

1. The need to increase the number of doctoral programs in life sciences and biotechnology, as a strong pool of academic leaders is essential to sustained innovation. A national working group agreed to establish programs of graduate and postgraduate model, to attract talent to the life sciences and provide working conditions for scientists to conduct industry-oriented research.

2. The need for more mature technologies such as diagnostics and vaccines. While the Indian industry is in strong product devel-opment and marketing business, biotechnology in India does not have the necessary infrastructure for R&D of molecular modeling, protein engineering, drug design, and immunological studies. The DBT[67] will act to facilitate a unique mechanism for

[64] DBT is a wing of Ministry of Science and Technology (MST), Government of India.

[65] There has been funding and assistance proposed for the biotechnology-related education and research at various levels. For the industry and business, various incentives and benefits have been proposed. The endeavors of research in biotechnology have also been recommended for stronger protection.

[66] The basic ideology behind the policy could be understood by going through the vision and mission of the policy as well as the main feature of the policy.

[67] For further details, please visit the official website of the Department of Biotechnology, Government of India at www.dbtindia.nic.in (last accessed 29 January 2016).

biotechnology plans and encourage private sector participation in infrastructure development.

3. India's strategy should aim to increase the value of R&D and production of intellectual property rights, which would result in more trade values and fulfilling consumer needs. India needs to provide active support through incubator funds and various incentives, and increased focus on innovation and the ability to create a pipeline of products continuously. Clear government policies to promote innovation and market knowledge will drive the growth of the biotechnology sector.

4. Government support, tax incentives, and tax benefits are essential for this sector. Biotechnology is the most research-intensive field and companies in the sector invest 20–30 percent of their operating costs in R&D or technology outsourcing. In addition, financial support for the development of early-stage products and small and medium-sized companies is the key to sustaining innovation.[68]

5. Creating a Small Business Innovation Research Initiative (SBIR), DBT, to support small and medium-sized enterprises through loans and grants. The support system for pre-proof of concept, innovation, and original research on mentoring.

6. Establishment of biotechnology parks to provide a mechanism for the licensing of new technologies for biotechnology companies to come to start new businesses and realize the value of early-stage technology with a minimum of financial data. Parks to facilitate technology transfer by serving as a stimulus for entrepreneurship through partnership among innovators from academia, R&D institutions, and industry. A number of state governments in India have shown keen interests in the concept of biotechnology parks.[69]

7. Need a mechanism for scientific, rigorous, transparent, consistent, and effective assessment of biosecurity; a single national

---

[68] For further details, please visit the official website of the Ministry of Science and Technology, Government of India at http://dst.gov.in/ (last accessed 29 January 2016).

[69] Chaturvedi and Rao. 2004. *Biotechnology and Development*, p. 139.

authority on biotechnology to be established and governed by an independent administrative structure.

Regulatory Mechanisms[70]

It is important that biotechnology is used for the social benefit and economic development of India. To fulfill this vision, it has to be ensured that research and application in biotechnology is guided by a process of decision-making that safeguards both human health and the environment with adherence to the highest ethical standards. There is consensus that existing legislation, backed by scientific assessment procedures, clearly articulates rules and regulations that can efficiently fulfill this vision. Choices that reflect an adequate balance between trade benefit, safety, access, and the interest of consumers and farmers need to be made. It is also important that biotechnology products that are required for social and economic good are produced speedily and at a lowest cost. A scientific, rigorous, transparent, efficient, predictable, and consistent regulatory mechanism for biosafety evaluation and release system or protocol is essential to achieve these multiple goals. The current biotechnology policy proposes sectoral road maps for the various segments of biotechnology research and development for the sake of overall development of biotechnology in the different sectors. Special attention has been given to agricultural biotechnology, industrial biotechnology, and medical biotechnology. At the same time, IT-enabled biotechnology, nanotechnology, and other such ultra-modern technologies and their development have also been mentioned quite comprehensively. The various sectors of biotechnology and the use of modern biotechnology for intellectual property creation, biotechnology based trade, commerce, consumable goods need to be regulated. At the same time, safety of environment and biodiversity along with securing public policy and welfare of the society need to be ensured during the application of biotechnology, for sustainable development.

---

[70] Regulation of any field or technology, for that matter, is required because anything that is left unregulated becomes either abused, misused, or overused, leaving bad effects of such field or technology for the future generations.

Biotechnology Policies of Developing Countries

Biotechnology has been subject of major policy considerations in the last two to three decades. The United Nations Conference on Environment and Development[71] asserted that biotechnology promises[72] to make a significant contribution in enabling the development of better health-care, enhanced food security through sustainable agricultural practices, improved supplies of potable water, more efficient industrial development process for transforming raw materials, support for sustainable methods of forestation and reforestation, and detoxification of hazardous wastes.[73] In the aftermath of the effort of the UN, a number of developed and developing countries endeavored to formulate national policies on biotechnology. At the beginning of the twenty-first century, the Republic of Korea under its Ministry of Science and Technology devised plans to set up biotechnology-related venture firms by helping entrepreneurs, scientists, and technicians. More than USD 250 million was earmarked for research and development in biotechnology. The major focus of the policy of the Korean government was to develop capacity in the areas of DNA, protein, and bioinformation. Through establishing the National Genetic Research Center, it intended to start the groundwork for genome studies and to maintain ethical order through introducing "life science ethics law" in the curriculum of genetic research.

The Nigerian government intended to promote biotechnology by investing, every year, USD 19 million on biotechnology and setting up the National Biotechnology Development Agency. The government policy plans to promote commercial utilization of biotechnology by the production of biotechnology products and services.[74]

In the meantime, Singapore announced the National Biomedical Science Strategy in 2000, to provide for an estimated investment of

[71] United Nations Conference on Environment and Development took place in the year 1992 with an ambition to integrate development with environment protection.

[72] Ricardo Melendez and Sanchez. 2007. *Trading in Genes*, p. 3.

[73] *Report of the United Nations*, 1992, p. 136.

[74] CorBiotech Update, 13 September 2002.

USD 2 billion for research, training, and entrepreneurship in bio-technology. In its policy, the government of Singapore announced tax incentives for entrepreneurs and the establishment and promotion of research institutes.

Post-2000, South Africa designed a national biotechnology strategy to spearhead the progress of biotechnology research. It intended to establish a government agency to promote and develop biotechnology through proactive investment in the field for human resources development and for the strengthening of scientific and technological capabilities.[75]

At the same time, Malaysia, an aspirant to become the hub of bio-technology, announced its biotechnology policy. Banking on rich and diversified bioresources, the Government of Malaysia wanted to use biotechnology to support sustainable growth. It introduced "BioValley Malaysia," a flagship program for biotechnology development in Malaysia for promoting research, industry, entrepreneurship, and networking in biotechnology.[76]

Having recognized the potential utility of biotechnology, the develop-ing nations wanted to commercially exploit it in the best possible way. The biotechnology policies of the developing nations discussed here highlight how these countries would like to take the support of biotech-nology for sustainable development of their economy and, at large, their nations. The policy showed how biotechnology was aimed to be used for research and development, industry and entrepreneurship, training and networking, capacity building, and human resources.[77] These nations were also aware of the complexities of biotechnology and the major areas of concern, both morally and legally. Therefore, they intended to include issues pertinent to social and legal order in the biotechnology strategy and policy to provide orientation with reference to intended regulation and code of conduct for biotechnology promotion and commercial exploitation.

[75] See Ricardo Melendez and Sanchez. 2007. *Trading in Genes*, p. 21.

[76] See also Chaturvedi and Rao. 2004. *Biotechnology and Development*, p. 143–50.

[77] See Ricardo Melendez and Sanchez. 2007. *Trading in Genes*, p. 21.

Technology Law

Technological revolutions in the modern world have brought several regulatory issues to the forefront. These issues include the promotion and protection of technological innovations; issues related to application and implications of technology to the growing needs of the society for faster, efficient, and durable results; regulation and sustenance of the use of such innovations; societal, ethical, moral, and religious concerns on such technological revolutions; use of such technological innovations, trade and commerce initiatives; liability and contractual concerns in dealing with technological revolutions and innovations; and so on.[78] Technology law addresses these concerns aiming at overall regulation and governance of promotion, sustenance, and dealings with technology, its innovations, applications, and implications. It is a new segment of law within the branch of business law. Defining the term technology law is difficult; technology issues touch almost all facets and branches of law, from general corporate law used to form and operate new and emerging companies that create and use technology to specialized disciplines such as intellectual property law, trade law, consumer law, and tax law. Even there are certain issues concerned with the technology that touches environmental law. In particular, use, transfer, and storage of biotechnology and nanotechnology innovations involve concerns about the environment and biological diversity. Besides, technology touches upon contract law and tort law in terms of mutual obligations of the parties involved with the dealings in technology, in terms of liability that arises out of use and storage of technology resulting in some damage or loss to the neighbor, giving rise to issues of negligence, strict liability, absolute liability, product liability, and so forth. Again, these concerns are felt more in case of dealings with biotechnology and nanotechnology innovations.

The law and technology have radically affected many aspects of people's lives: how we work, how we learn, how we spend our leisure time, how we communicate, and how we do business. The scope of technology

---

[78] See Sreenivasulu N.S. 2013. Law and Policy of Science and Technology in India, A policy document and report submitted to the MHRD, Government of India through the funding of Distance Education Council and Karnataka State Open University, Mysore.

is very broad; it affects so many aspects of law in both personal and commercial dealings, making it virtually impossible to define technology law by substance any more.[79] Technology law is not a discrete area of law but instead draws itself into many different legal arenas. There is an assumption that technological revolutions and innovations will continue to challenge the cantors and cannons of our traditional legal framework. The challenges posed by technology have made us reexamine and reassess the strategies to deal with those technologies.[80]

Major Components of Technology Law

Technology has become an important part of the lives of people and it is an essential tool for spearheading artificial entities such as business enterprises. In the contemporary world, people in their day-to-day lives and entities in their day-to-day affairs use and deal with technology. In the present technological world, there cannot be any person or entity not affected by technology or not using its products directly or indirectly, thus commercializing and commoditizing technology and its applications. Technology law, though not an independent area of law as such, addresses the legal issues arising out of the commercialization of technology. It deals with various issues ranging from promotion and progress of technology, industrial application and commercialization of technology, dealings in technology, consumer issues, safety, sustenance, and secured use of technology,[81] transfer of technology to investment, trade, and competition in technology business. The promotion and protection of any technology can be done by providing incentives for the innovators through Intellectual property law that provides for incentives in the form of exclusive monopoly to the inventors and authors making intellectual property law a major component of technology law. In many ways, technology law is synonymous with intellectual property law, as

---

[79] Aspatore. 2004. *Inside the Minds: The Laws Behind Technology: Leading Lawyers on the Legal Aspects of Patents, Software Licensing, Telecommunications & More*, p. 38. USA: Aspatore Books.

[80] Aspatore. 2004. *Inside the Minds* p. 58.

[81] See generally Ricardo Melendez and Sanchez. 2007. *Trading in Genes*, p. 153.

technology is knowledge, which, in turn, is an intangible asset: the basic trait of intellectual property.[82] Transfer of technology from one person to others is a commercial activity bringing into picture commercial law and contract law, which generally govern mutual contracts regarding the use of technology, acquisition, and sale of technology. Technologies such as biotechnology and nanotechnology involve the release of nonnatural living beings into the environment and, considering the question of safety of such release, environmental law and biodiversity laws come into play. Use and supply of technology and its products involve liability concerns, giving rise to application of product liability and consumer laws. Trade and business in technology brings into play trade and investment law as well as competition law. Safety, security, and sustenance issues in technology would bring into play environment and biodiversity laws. At the outset, it could be said that technology law is a manifested law that includes components from various laws based on the varied applications of respective technology.

Biotechnology Law

There is no categorical law such as biotechnology law. It is a culmination of rules, regulations, and laws pertinent to regulation of various issues and concerns involved in biotechnology innovation, development, and use. Biotechnology law basically deals with the issues concerning promotion, protection, and enforcement of biotechnology innovations, biotechnology trade and business, regulations and safety of its use and sustenance, societal, ethical, and religious concerns, consumer interests, and implications of tort law, human rights law, contract law, corporate law, and other such branches of business law and public law on biotechnology. When it comes to the question of regulation of biotechnology, probably the most complicated technology at hand, the issues are concerned not only with conventional corporate laws and intellectual property laws[83] but with environmental and biodiversity laws. Since biotechnology deals with living beings and creates nonnatural living beings, the impact on

---

[82] Furthermore, the law of intellectual property rights is used to analyze who owns what rights to technology, making it more or less the law technology.

[83] See generally, Sreenivasulu. 2008. *Biotechnology and Patent Law.*

the environment and the biological diversity needs to be checked.[84] Nonnatural livings beings and their use, storage, transfer, and release into the environment involve taking certain safety measures to avoid any damage to the environment. Improper use and release of biotechnology and nanotechnology[85] innovations might cause damage to the environment and ecology. We have seen controversies pertinent to biotechnology in the cases of oncomouse, Dolly sheep, beef hormone, monarch butterfly, EC biotech, and so forth, in the US, Europe, and such other developed nations. There have been many controversies over genetically modified crops in India, for instance, as seen in the case of Bt Cotton (cash crop) and Bt Brinjal (food crop). These case studies have also revealed that the current legal and regulatory framework with respect to biotechnology is inadequate. Nations such as the US and the UK could bring in some amount of legal mechanism and policy initiatives with reference to biotechnology. The developing nations are skeptical with reference to the fabric and content of the proposed legal frameworks in this regard. The law will have to address a wide range of issues concerning trade, environment, intellectual property, consumer protection, human rights, and ethical, social, and religious issues at both international and national levels.[86] While formulating policies and implementing laws, the approach should be of balancing public and private interests with sustainable development. However, it is felt that besides law and policy initiatives, much is required to be done on the infrastructural front and capacity building in biotechnology. There are a number of issues and concerns that need to be addressed in the biotechnology policy formulation and legal regulation, including the following.

- Promotion of innovation and development for the larger needs of the people and greater benefit of the society.
- Protection of research and innovation for promoting the commercialization of innovations to boost the economy.

[84] See generally, Ricardo Melendez and Sanchez. 2007. *Trading in Genes*, p. 153.

[85] Naidu. 2009. *Biotechnology and Nanotechnology*, p. 33.

[86] See also, Ricardo Melendez and Sanchez. 2007. *Trading in Genes*, pp. 11–15.

- Incentivizing creative endeavors through protective mechanism and exclusive monopolies on innovations.
- Right to know/information about the technological advancements and scientific developments in the public interest.
- Inventor's right on the innovation, confidentiality, and monopoly through the grant of intellectual property rights.
- Balance of interest: Conditional temporary monopoly and compulsory disclosure for maintaining the equilibrium in terms of knowledge exploitation and use.
- Promotion and regulation of trade-related aspects of biotechnology including trade in GMO/LMO and adhering to international obligations.
- Issues concerning clinical trials[87] and marketing approvals for biotechnology products and GMO.
- Consumer concerns in trade in biotechnology and GMO, labeling, accessibility, and suitability of GM food.
- Technology development and transfer including development of indigenous technology and cross-border transfers and related issues.
- Human Rights approach to biotechnology research and development: Issues concerning integrity and dignity of human beings, ensuring right to food[88] and right to health through biotechnology.
- Ensuring biosafety in GMO creation, trade, and development, while securing the health and safety of humans and other living beings.
- Regulation of access to genetic and biological resources and recognizing the rights of indigenous communities, obtaining their consent, and sharing of benefits.

---

[87] Government of India imposed moratorium on clinical trials of at least 15 GM crops on 29 July 2014. These crops were earlier cleared for clinical trials by the Genetic Engineering Approval Committee functioning under the Ministry of Environment and Forests.

[88] Government of India enacted National Food Security Act, 2013, to ensure right to food to the people. It is far widely recognized that the objective of ensuring right food cannot be achieved without promoting and using biotechnology in Indian agriculture.

- Protection of biodiversity, prohibiting biopiracy and facilitating sustainable development for the benefit of mankind.
- Addressing public policy issues and societal concerns in biotechnology research including social acceptance of biotechnology, its use, and development.

Indian Policy Response to Biotechnology Upsurge and Regulation

India seems to be one of the few developing countries in the world to perceive the importance of biotechnology much before it became an established industry even in the developed countries. Its response to the biotechnology rush has been quite satisfactory in terms of realizing its crucial potential as well as attempting to realize the possible risks associated with it. A look at the various attempts of the Government of India in putting into place a number of policy and legislative initiatives as well as the establishment of diverse means and mechanisms would give an indication that India is not really lagging behind in handling the biotechnology upsurge.

The Indian government established the National Biotechnology Promotion Board in 1982, which, in 1985, was converted into the Department of Biotechnology (DBT) under the Ministry of Science and Technology, one of the earliest such departments in the world.[89] In accordance with the powers conferred on the Government of India under the Environment Protection Act of 1986, it promulgated the Rules and Procedures for the Manufacture, Import, Use, Research, and Release of Genetically Modified Organisms in 1989, which was amended in 2006.[90] The aim of these rules is to ensure that the use of genetically modified products is safe to environment and beneficial for human beings. These rules have been revised in 1994 (after India signed the CBD Convention) and in 1998. They constitute the basic minimum requirements for those involved in research in biotechnology

[89] See www.dbtindia.nic.in (last accessed 3 February 2016).
[90] 'Rules for manufacture, import, use, research, and release of genetically modified organisms'. 1989. Ministry of Environment and Forests, available at http://www.moef.nic.in/legis/hsm/616E.pdf (last accessed 2 February 2016).

in India. The DNA safety guidelines were brought into the regulatory framework in 1994. A multi-tier regulatory system has been brought into place in India for biotechnology regulation. India follows a three-tiered system for maintaining biosafety. The first tier in this regard is the Institutional Biosafety Committee (IBSC), which oversees more than 160 institutional biosafety committees throughout the country including all research organizations, companies, and universities engaged in transgenic research. The second tier is the Review Committee on Genetic Manipulation (RCGM), which gives the clearance for conducting clinical trials in ex situ and in situ environments. The RCGM is a monitoring cum evaluation committee that visits the sites, collects data, and evaluates it for risk factors. The third tier is the Genetic Engineering Approval Committee (GEAC) under the MoEF, which grants permission for large-scale field trials.[91] The last chapter of this book is dedicated to discussion on "Biotechnology Regulation in India," with a detailed analysis made on a number of guidelines, regulations, rules, norms, and institutional mechanism related to biotechnology regulation.

The Patent Act was amended in 1999, 2002, and 2005 to provide for patenting of genetically modified microbes and other such biotechnology innovations. There is a detailed discussion on intellectual property and patent law pertinent to biotechnology in the next chapter. The Plant Varieties Act and the Biodiversity Act were brought into place to protect traditional knowledge and biodiversity in India, to provide for biosafety mechanism in the backdrop of biotechnology upsurge while providing for bioprospecting leading to genetic innovations. Under the mandate of the act, the National Bio-resource Development Board under the aegis of Department of Biotechnology has developed an information technology database known as "Jeev Sampada." The database prepared through collaboration with several institutes and experts is a digital documentation of information on plants, animals, and marine and microbial resources of India. The Biodiversity Act that deals with "trade environment-related aspects of biotechnology" has been discussed in detail in Chapter 5.

The Food Safety and Standards Act, 2006, that replaces the earlier food safety act, brings in the sensitized regulatory framework for

---

[91] Detailed description of the duties and functions of the committees can be found at MoEF's website: www.envfor.nic.in (last accessed 3 February 2016).

regulating GM foods safety while adopting the guidelines of International Codex Alimentarius on food quality and standards.[92] It is the primary legislation among others to ensure that the GM food produced is fit for human consumption. The Department of Consumer Affairs through a gazette notification has made it mandatory to label the food containing GM organisms from 1 January 2013. In 2014, the Government of India brought in National Food Security Act to provide food security to the masses. The act acknowledges the contribution of biotechnology to increasing agricultural yield and food production. Chapter 6 in this volume that deals with "biotechnology and human rights" also discusses the aforementioned two legislations while presenting a case for possibility of ensuring right to food and food security through biotechnology.

The need for an integrated biotechnology policy with concurrent attention to education, social mobilization, and regulation is considered to be an essential prerequisite for an orderly progress of the biotechnology sector. Synergy between technology and public policy is essential for us to achieve an effective mobilization of the tools of new biology to add both years to life and life to years. The biotechnology policies[93] of the nations shall take into consideration these issues and concerns, including international policy options and obligations[94] that will affect the biotechnology industry. There could be multilateral and bilateral understandings for mutually promoting, respecting, and cooperating in biotechnology endeavors. The policies shall chalk out directions to strengthen the nation's industrial biotechnology research capabilities; work with business, government, and academia to move biotechnology from research to commercialization; foster the nation's industrial development; inform people about the science, applications, benefits, and issues of biotechnology; and enhance the teaching and workforce training capabilities. In today's context, it is imperative that India leverage resources through partnership and build regional innovation systems. The strategy will help develop local talent for a globally competitive

---

[92] See User's Manual on Codex—India (FAO), available at http://dbtbiosafety.nic.in/act/codex.pdf (last accessed 3 February 2016).

[93] The Government of India brought in National Biotechnology strategy, which is an exclusive policy framework for promoting biotechnology in India.

[94] See Ricardo Melendez and Sanchez. 2007. *Trading in Genes*, p. 11.

workforce. While recognizing private sector as a crucial player, the nations shall visualize in their strategies the government to play a major catalyzing role in promoting biotechnology. The Indian biotechnology development strategy is based on a strong innovation promotion framework in which industry, academia, civil society organizations, and regulatory authorities will communicate in a seamless continuum. It is necessary that the perspective for Indian biotechnology law and policy shall be global and at the same time would address the local concerns.

4   Intellectual Property Rights in Biotechnology

Human intellect and creativity, if applied in the field of biotechnology, would result in a number of innovations and creations. Biotechnology is also used as a popular term for the genetic technology of the twenty-first century. Although it has been utilized for centuries in traditional production processes, modern biotechnology is only about 50 years old, and in the last decades, it has witnessed tremendous developments. One of the most important issues that has been raised due to the emergence of modern biotechnology is the legal characterization and treatment of intellectual creations described as intellectual property. Technological innovations are usually protected by intellectual property rights (IPRs)[1]—a topic that has been the subject of discussion in recent years.[2] In this connection, one may like to compare

---

[1] Wuger, Daniel and Cottier Thomas. 2008. *Genetic Engineering and the World Trade System: World Trade Forum*, p. 110. Cambridge, UK: Cambridge University Press.

[2] See generally Sreenivasulu, N.S. 2011. *Intellectual Property Rights*, 2nd edn, p. 3. New Delhi: Regal Publications.

biotechnology with other technologies in which the advances are covered by the patent and intellectual property system and are, therefore, routinely licensed and marketed. To protect intellectual endeavors, which are time-consuming and require huge investments, resulting in novel innovations, and the commercial interests of the stakeholders, it is required to ensure protection of IPRs. There are, however, no internationally accepted guidelines for the management of IPRs, and a wide range of opinions exist regarding the utility of intellectual property in the area of biotechnology. However, there are different options available within the framework of intellectual property law for extending protection to the intellectual endeavors and innovations of biotechnology and genetic research. This chapter aims at presenting the development of intellectual property mechanism toward the protection of biotechnology innovations and exhibiting different options within the framework of intellectual property law, including but not limited to patents, trade secrets, and copyrights in different regions.

Biotechnology Innovation and Intellectual Property Protection

Many of the new biotechnological innovations are protected by IPRs, raising serious concerns about access and use of technology and resources.[3] The term "property" is often found associated with physical and tangible objects only, such as household goods or land, for which ownership and associated rights are guaranteed and protected by the law prevalent in a country. Intellectual property, on the contrary, is intangible and includes patents, trade secrets, copyrights, trademarks, geographical indications, industrial designs, and semiconductor chips.[4] In the field of biotechnology, intellectual property refers to the innovative processes and products that are an outcome of the development of genetic engineering techniques using restriction enzymes to create recombinant DNA (rDNA). The characterization of these researches

---

[3] See Ricardo Melendez, Ortiz and Vicente Sanchez. 2007. *Trading in Genes: Development Perspectives on Biotechnology, Trade and Sustainability,* 1st edn, p. 124, London: Earth Scan.

[4] The rights to protect this property prohibit others from making, copying, using, or selling the proprietary subject matter.

results as intellectual properties which encourages industries to allocate labor, research and development (R&D) units, and funding to facilitate the production of commercially marketable items. At the same time, IPRs for biotechnological inventions raise distinct sets of issues concerning and including development of and access to technology, benefit sharing, acknowledging the source, respecting traditional knowledge, issues of public domain, private monopoly over biological resources, bioprospecting, biopiracy, and sustainable development, to mention a few. Due to these intellectual properties, many legal policies, which are perhaps considered as impediments to biotechnological research, are also being challenged and are, therefore, undergoing tremendous changes. This is understandable because if public policies do not allow the development and commercial use of an intellectual property, no industry would like to invest funds in this research.[5] More recently, however, utility patents for genetic materials, both plants and animals, have also been allowed in some countries, so that neither the patented material can be used for further breeding nor will the farmers be allowed to save and use the seeds for cultivation, without paying a fee to the patent holder. Similarly, if patents on superior animal breeds are allowed, a dairy farmer will find that a calf born to his hybrid cow will belong to the company that sold him the cow. These IPRs affect the availability of genetic diversity as well as the conservation, availability, and use of plant and animal genetic resources. One of the key concerns in the area of IPRs and biotechnology is the patenting of research tools or the grant over broad patents that could potentially block further useful research. It is felt that there could be no access to information and technology for further study and research if there is any copyright protection. Similarly, trade secret protection to biotechnology would keep the innovation secret from and inaccessible to the public. There are also arguments against patenting life forms such as transgenic animals and plants because these patents will work as impediments in free exchange of

---

[5] Another example of intellectual property is the development of crop varieties that are protected through 'plant breeder's rights' or PBRs. Through PBR, the plant breeder who has developed a variety enjoys the exclusive right for marketing the variety, although the use of the variety for further breeding or for replantation of seeds saved by a farmer (farmer's exemption) is permissible.

genetic materials for improvement of crops and livestock. It is believed that IPRs may also adversely affect

- food and health security;
- use of evolved agricultural practices;
- access to genetic information and innovations;
- biological diversity and ecological balance;
- livelihood of the poor in the developing countries.

Biotechnology contemplates modified innovations across different species including plants, animals, and microorganisms. Biotechnology also involves human genetic innovations including human genes, DNA, their produced proteins, enzymes, and so on. Similarly, biotechnology-assisted processes produce genetically modified plants, animals, and other such living products. Better understanding of intellectual property by research scientists and university administrators will increase the research in scientific and technological innovation and development of biotechnology. IPRs are indeed indispensable tools needed to foster the common concern associated with biodiversity and biotechnology.[6] Appropriation of biological and genetic resources through biotechnology is likely to be protected through patents, trade secrets, and other such viable options under intellectual property regime.

Intellectual Property with Reference to Biotechnology

Perhaps, the need for protection of IPRs stems from the developed nations where free flow of finance for research has resulted in progress of R&D. The more the research, the more the claims for IPRs and the more the demand for protection of the same through exclusive monopolies. Developed nations demand for strong intellectual property regimes across the globe. The developing nations, because of their own limitations and constraints, are neither able to provide for the promotion of research in science and technology nor able to generate claims for IPRs, and therefore not very keen on a strong intellectual property regime. The experiences in the developed nations state that the promotion of

---

[6] Francioni, Fransesco and Tullio Scovazzi (eds). 2006. *Biotechnology and International Law*, p. 36. Portland, USA: Hard Publishing.

science and technology with appropriate protection in the form of intellectual property regime contribute to the progress of the society. The development of capabilities for the effective management of intellectual property is an important element in securing the benefits of public and private sector research in biotechnology. In this context, filing of patents both in India and aboard are critical to the growth of the Indian biotechnology sector. The expenses for filing patents, especially outside India, are prohibitive and a major barrier to effective intellectual property management within the country. While expenses incurred with respect to filing of patents in India is eligible for weighted deduction, similar benefit is not provided for expenses incurred with regard to filing patents outside India.[7] This is also imperative in the new World Trade Organization (WTO)-Trade-Related Aspects of Intellectual Property Rights (TRIPS) regime, which has taken effect from 1 January 2005.[8] Administration of the new IPR regime should be improved. This will be achieved in the following ways:

- Encouraging science graduates to pursue law for better understanding of IPR-related issues.
- Including IPR-related issues in the curriculum of law colleges for facilitating filing of international patents, license negotiation, dispute resolution, and so forth.
- Training scientists and technology transfer professionals in the strategy of intellectual property protection related to assessment of patentability, prior art examination, and technology transfer issues.
- Training patent attorneys on science subject(s) and improving mechanisms for IPR administration through reforms and creation of patent offices, patent codes, and ensuring adequate availability of patent attorneys. This will be promoted through an effective interministerial collaboration.

[7] As IPR creation is a prerequisite for exports to the regulated markets, it is recommended that expenditure incurred with regard to filing patents outside India be also eligible for weighted deduction U/S 35 (2AB).

[8] India ratified WTO/TRIPS agreements in 1994 and the agreements have been implemented at various levels in India during the period of 1995 to 2005.

- Setting up of an arbitration council to redress IPR disputes, which will help in improving the perception and increasing international confidence toward IPR protection in India.
- Allocating a budget of Rs 50 crore to substantially improve the current patent infrastructure and set up additional offices in cities such as Bengaluru and Hyderabad.
- Constant dialogue between the Department of Biotechnology and the Government of India and WTO-TRIPS to address patentability issues in biotechnology and their future inclusion in the Patents Act through amendments.

The TRIPS Agreement, Biotechnology, and IPR

IPRs, since their inception, have been shaped by a number of international treaties. National legislators have had to study the international scene to gain some insight into the prevailing intellectual property standards.[9] It could be said that formal protection of IPRs in biotechnological innovations started with the adoption of the TRIPS agreement, as international intellectual property law led by the TRIPS prescribes obligations for the member states. The agreement provides for a common and uniform set of rules to protect and enforce IPRs. The purpose of the agreement is to reduce distortions and impediments to international trade, to promote effective and adequate protection of IPRs, and to ensure that measures and procedures to enforce IPRs do not themselves become barriers to legitimate trade. Opinions differ among nations about the protection of IPRs in biotechnology innovations; in particular, countries such as the US supported it, but nations such as the European Union (EU), Canada, and India were against or not very much in favour of it. At the time of the negotiations on TRIPS in the Uruguay Round held during 1986–94, the US and the EU differed in their approaches to the protection of biotechnology innovations, in particular regarding patenting. While the US believed that "anything under the sun made by man," except for human beings themselves, was patentable, the EU was grappling with strong internal resistance to intellectual property

[9] Wuger and Cottier. 2008. *Genetic Engineering and the World Trade System*, p. 77.

protection of biotechnological innovations and living organisms.[10] It is believed by the biotechnology industry that there is blanket permission for biotechnological innovations for patents under the TRIPS agreement when it stated that patents shall be made available to innovations in all the fields of technology, including biotechnological inventions. Article 27(3)(b) of the TRIPS agreement provides flexibility[11] to member states to exclude[12] plants and animals other than microorganisms and biological processes for the production of plants or animals other than microbiological processes from patent protection. Most of the developing nations, including India, have used the flexibility and excluded plants, animals, and biological processes from the purview of their domestic patent laws. It is quite relevant to mention here that Article 27.3(b) of TRIPS excludes biological processes for the production of plants or animals from the patentable subject matter. This particular provision of the TRIPS agreement is very significant in the context of biotechnology industry that depends upon biological resources for the development of new and innovative genetically modified biotechnological inventions. The agreement seemingly recognizes the patentability of biotechnological inventions and the enjoyment of rights thereof.[13] International patent[14] laws encourage using microorganisms and nonbiological and microbiological processes in the production of nonnatural plants and animals. Under the Indian Patent Act, similar impression has been given regarding microorganisms and

[10] Ricardo Melendez and Sanchez. 2007. *Trading in Genes*, p. 124.

[11] TRIPS agreement provides option to member states for protecting new plant varieties by means of patent or sui generis system or both. India opted for sui generis protection and legislated "Plant Varieties Protection and Farmers Right Act-2000" that enabled the farmer to save, use, sow, re-sow, exchange, or share the seeds of protected variety, besides offering protection on farmers' variety, extant variety, and essentially derived variety.

[12] Wuger and Cottier. 2008. *Genetic Engineering and the World Trade System*, p. 79.

[13] Francioni and Scovazzi (eds). 2006. *Biotechnology and International Law*, p. 42.

[14] Indian Patent Act 1970 defines patentable invention as a new product or process involving an inventive step and capable of industrial application. See Sreenivasulu. 2011. *Intellectual Property Rights*, p. 30.

nonbiological and microbiological processes. This extension of patent cover includes even the gene sequences coding for any particular character, a promoter or genetic markers, or similar ones. With the increase in transgenic research in both public and private research organizations, the issues of royalty payments, material transfer agreements (MTAs), and legal obligations and bindings need to be clearly understood. Nevertheless, IPR protection of new life forms of biotechnology raises a number of difficult technical and ethical issues because of which the patentability of new biological forms and processes is still not accepted in many countries. Since intellectual property protection is granted only for inventions and not for discoveries, in case of biotechnology innovations it is difficult to say whether the new life form in the form of gene, DNA, cell, etc. is a scientific discovery or a technological invention. However, in a few countries, such as the US, even discoveries are protected under patent law without any distinction.[15] Discovery is considered to be merely making available what already exists in nature. A substance freely occurring in nature, if merely found or discovered, is not patentable under most of the jurisdictions, including India and Europe. However, if the substance found in nature has to be first isolated from its surroundings with a specialized process, that process is considered an invention and hence is patentable. But in India, there are several ethical issues related to patenting of life forms, the most important being the issue of extent of private ownership that could get extended to the life forms. One of the major causes of uncertainties and controversies related to IPR protection of life forms is lack of an established practice to protect not only such living materials, but also any form of intellectual property. However, it is not only about ownership of physical property that has a tangible market value; in the traditional cultural context, Indians have considerable problems fixing monetary value to anything that is not a tangible physical property. Hence, there is an urgent need for developing countries like India to define clear policies for IPR in case of scientific and technological innovations. The idea of profit-making by exploiting any common heritage of civilization or culture is abhorrent to many people and communities. In the case of modern biotechnology,

---

[15] Sreenivasulu, N.S. 2008. *Biotechnology and Patent Law: Patenting Living Beings*, 1st edn, p. 64. Noida: Manupatra Publications.

these objectors do not allow much innovation, and therefore argue that any form of IPR must not be granted. Although it is believed that pro-tection of innovation in plant variety could boost research in the area of plant biotechnology by both public and private bodies, it could also result in higher prices for seeds, thus naturally excluding the small and marginal farmers from accessing such new technologies. Farmers and indigenous people in developing countries such as India are facing seri-ous problems, as plants that they developed and conserved are being "appropriated" by private entities leading to biopiracy and exploitation of traditional knowledge claiming the exclusive right to produce and sell many "modified" plants and animals. This is a matter of great concern today that knowledge, innovation, and efforts of these communities are not acknowledged when the legal "intellectual property rights" systems grant patents on genetic and biological materials and living organisms to private corporations. In 2000, the Council for Scientific and Industrial Research (CSIR) found that almost 80 percent of the 4,896 references to individual plant-based medicinal patents in the United States Patents Office that year related to just seven medicinal plants of Indian origin. Three years later, there were almost 15,000 patents on such medicines spread over the US, the UK, and other registers of patent offices. In 2005, this number had grown to 35,000, which clearly demonstrates the interest of the developed world in the knowledge of the developing countries. However, during the same time, a number of patents were granted by the US and other patent offices on Indian biological resources and traditional knowledge. These controversial patents included patents to American companies on modified varieties of "basmati," "turmeric," and "neem," which involve credible Indian traditional knowledge and biological resources.[16] There have been a number of such instances where patents on Indian biological resources were obtained by multina-tional companies operating across the world.[17] While the corporations

[16] Sreenivasulu, N.S. and K.S. Kariyanna. 2012. 'Traditional Knowledge and Intellectual Property Rights', *Manupatra Intellectual Property Rights*, I (I): F-17.

[17] Sreenivasulu, N.S. and K.S. Kariayanna. 2012. 'Regulation of Biological Resources and Biodiversity', *Manupatra Intellectual Property Rights*, I (II): F-117.

stand to make huge revenues from this process, the local communities remain unrewarded and, in fact, face the threat in future of having to buy the products of these companies at high prices. Hence, it is argued that such system of IPR only benefits the private industries or multinational corporations of industrially developed countries at the expense of the developing countries. There is a need to define guidelines and policies for the implementation of IPR so that people, like farmers, get recognition for their efforts and contributions to prevent biopiracy.[18] Several civil society organizations (CSOs) and nongovernmental organizations (NGOs) argue that naturally occurring organisms are God's gift and nature's bounty; they are common property of the mankind, and therefore cannot be appropriated by any person(s) or organizations or entities by just modifying it or tinkering with it. It is widely felt that a number of NGOs from across the globe, including from the developed countries, are raising the concerns of the developing nations in IPRs under the TRIPS agreement.[19]

Biotechnological Innovations

In general, biotechnological innovations could be classified as follows, protected within the broader purview of intellectual property law:

1. genetic sequences of living beings and genetically modified living organisms;
2. genetic databases and information related to the genetic identity and structure;
3. DNA sequences and their structure;
4. biotechnological information and bioinformatics;
5. literature related to biotechnological and genetic innovations;
6. processes involving the invention of new genetically engineered microorganisms, plants, or animals;
7. process to genetically engineer microorganisms, plants, or animals;[20]

[18] World Intellectual Property Organization (WIPO) is now developing guidelines to protect traditional and indigenous knowledge systems.

[19] See Ricardo Melendez and Sanchez. 2007. *Trading in Genes*, p. 134.

[20] See Sreenivasulu. 2008. *Biotechnology and Patent Law*, p. 16.

8. processes of isolating genes, DNA gene fragments, and cells from one living being and incorporating them into any other living being;
9. processes of deriving proteins through expression of a foreign gene in a host organism;
10. process or method of treating genetic or hereditary diseases, known as gene therapy;
11. process of fermentation of beverages and cheese;
12. process to clean up pollutants in the environment;
13. genetically modified livestock, including fisheries;
14. gene therapy and genetic detection and surgical methods;
15. cells including stem cells;
16. genetically modified or engineered organisms (GMO), including microorganisms, plants, or animals;[21]
17. isolated and purified cells, genes, DNA, or gene fragments from the body of living beings;
18. proteins and enzymes identified in the body of living beings and commercially produced outside the body through industrial processes;
19. genetically engineered food and beverages, including genetically modified and consumable food;
20. development of genetically modified crop and plant varieties.[22]

These innovations could be protected under intellectual property law including patents, copyrights, trade secrets, and plant varieties law. Depending upon the nature, application, and compatibility, innovators can choose the right sphere of intellectual property for protection of identified innovations of biotechnology.

Choice of Intellectual Property Protection

Patents, trade secrets, and copyrights can be bought, sold, and licensed like any other property. Of course, making choice of an appropriate

[21] Wuger and Cottier. 2008. *Genetic Engineering and the World Trade System*, p. 78.

[22] Sreenivasulu, N.S. 2006. 'Protection of Plant Varieties and Farmers Rights', *Journal of World Intellectual Property Rights*, Serials Publications, I.

mode of protection is a business judgment. It depends upon the nature of innovation and the subject matter of protection. The choice of protection also depends on the nature of subject matter that needs to be protected. For instance, an instruction manual can either be copyrighted or protected as trade secret, but cannot be protected as a patent. The choice of protection is based on several factors that depend and differ from case to case. These factors include the following:

- Pace of technological development: If the technological development concerning the innovation is rapid, then a trade secret approach may be preferable to patenting.
- Associated costs: Although the cost of secrecy may exceed the cost of obtaining a patent, it may be more expensive to enforce the patent; so one will have to decide whether trade secret is a better protection.
- Images/Sketches: In case of genes and DNA, literature regarding sequencing and actual sketches and images of the invention could be protected under copyright before commercially exploiting the same.
- Security considerations: It may be impossible to prevent disclosure of a trade secret, despite all the precautions taken by the owner. So, in such cases one may prefer patent rights other than secrecy.
- Need to show patents: Patents may have to be shown to investors as a measure of success. For attracting investments, patents would garner more weight than any other IPR.

Trade Secrets Protection to Biotechnology

In the discussion of IPRs, trade secret protection is often overlooked. In cases where there is little possibility of independent discovery and innovation, and where the innovation is questioned for being independent from its natural counterpart or not, trade secret protection could be a viable option. Since there is no innovation requirement in trade secret, any secret and sensitive information concerning the possible or projected innovation or the innovation itself could be protected. Trade secrets often include private, sensitive, commercial, and proprietary information or physical material that allows a definite advantage and competitive edge to the owner.[23] This can be illustrated by the popular example of Coca-Cola brand syrup formula, which is well protected and popular trade secret.[24] Trade secrets in the area of biotechnology may include material such as:

- amino acid sequences of genes;
- genetic structure of living beings;
- hybridization conditions;
- cell lines;
- biotechnological methods or processes;
- method of isolating genes and DNA;
- gene therapy;
- formula to express foreign genes;
- formula to suppress the activities of genes;
- corporate merchandising plans of biotechnology;
- customer lists.

Unlike patents, trade secrets have an unlimited duration and may not be required to meet the conditions like the ones laid down by law for patent applications. Disclosure of a trade secret and its unauthorized use can be punished by the court and the owner may be allowed compensation. However, if a trade secret becomes public knowledge by independent discovery or other means, it is no longer protectable. Due to a large degree of research component in biotechnology, there is increased risk in maintaining trade secrets. The research results are published and discussed in conferences, and disclosed by exchange among graduate students, thus making it difficult to maintain secrecy. Despite this, in some cases, reliance on trade secrets is more prudent than the patents, which sometimes may become outdated before the patent is granted.[25] According to TRIPS agreement under Article 39, trade secrets have to

---

[23] Sreenivasulu, N.S. 2013. *Law Relating to Intellectual Property*, 1st edn. Bloomington, IL, USA: Penguin-Partridge Publications.

[24] Sreenivasulu. 2011. *Intellectual Property Rights*.

[25] It takes about approximately 1.5 to 2 years in the US to get a patent counting from the date of filing the patent application. It takes 2–3 years in Europe and 3–4 years in India for the same.

be protected in all the WTO countries. [26] Trade secrets protect against misuse or misappropriation of inventions that are kept secret, and not against independent discovery or reverse engineering. An invention is therefore a good subject matter for trade secret protection when:

- the invention embodies a high degree of complexity and novelty that would make independent invention by a competitor unlikely;
- the novel aspects of the invention are not embodied in a form that would permit the invention to be reverse engineered; and
- the invention is so securely protected that it is not likely to walk out of the door with a customer or an ex-employee.

The classic example of a good candidate for trade secret protection is a breakthrough manufacturing process that cannot be deduced from the end product. For example, a discovery that a previously patented drug can be synthesized more efficiently using an unusual set of reagents and reaction conditions could be preserved as a trade secret.[27] Patenting of such a process would require disclosure of the invention to the public, and the risk of infringement might not provide sufficient deterrent to prevent others from secretly using the process. Nevertheless, such an invention should be maintained as a trade secret only if the company is committed to taking the stringent security measures necessary to maintain the secrecy of the invention. In determining whether to maintain an invention as a trade secret or not, it is important not to stumble over the "best mode" requirement. Patent applicants often like the idea of patenting a basic invention while concealing detailed information about some special aspect of the invention, such as a superior method of making it. However, the patent law requires the patent application to describe the best mode subjectively known to the inventor for the invention. Failure to include the best mode can result in the invalidation of the resulting patent; thus, patent applicants should not attempt to retain the best mode as a trade secret while seeking a patent on other, that is, the less effective modes of carrying out the invention. In some circumstances, an

---

[26] Ricardo Melendez and Sanchez. 2007. *Trading in Genes*, p. 129.

[27] See generally, Sreenivasulu, N.S. 2012. 'Trade Secrets', in Paramjit S. Jaswal, G.I.S. Sandhu, and Anand Pawar (eds), *Consumer Activism, Competition and Consumer Protection*. Patiala, India: Rajiv Gandhi National University of Law.

opportunity may exist to patent certain aspects of an invention while retaining other aspects as trade secrets.[28] However, usually trade secrets are used to supplement and not substitute other IPRs, as sometimes protection under trade secret regime may not be as strong as it appears. If the invention is not a good candidate for trade secret protection, the next step is to consider whether patent protection is warranted.

Trademarks and Biotechnology

Although there could be no direct link between the law of trademark and law of biotechnology, there are some concerns in the use and regulation of biotechnology, which needs treatment based on the principles of trademark law. Labeling is required for genetically modified organisms, in particular genetically modified food. Nations like the US do not prescribe mandatory labeling, but regions such as the EU and India prescribe for mandatory labeling.[29] Serving the main function of trademarks to enable the consumers to identify and distinguish the goods and services on which the trademark is used, it can be said that trademarks enable the consumer to distinguish the genetically modified food from non-genetically modified food. There are practices of using a particular sign or symbol connected with biotechnology industry, in particular with the innovation that is being used as a trademark for representing its enterprise through commercialization. This practice is widely seen in the biopharma industry, where the name of the disease, drug, vaccine, or medicine is being used as a trade name or trademark to indicate to the customer that the product is biopharma and produced through genetic engineering. Likewise there are reflections of trademark propriety in the biotechnology innovations and their commercialization. Since under the trademark law and practice, it is very much established that there shall be no prohibition on use of names as business identifiers and trademarks, there shall be equally no restrictions on using the name of actual innovation, its ingredients or components, name of the

---

[28] However, decisions about such approaches should be made in light of a comprehensive strategy in consultation with a patent attorney or agent who has been fully informed on all aspects of the invention.

[29] Evenson, R.E. and Terri Raney. 2007. *Political Economy of Genetically Modified Organisms*, p. 221. Glasgow, UK: Edward Elgar Publishing.

drug, vaccine, the disease or organ that the drug, medicine, or vaccine treats, and any such signs, titles, or symbols that are directly or indirectly connected with an identified biotechnology innovation. Similarly, any sign or symbol used for representing biotechnology goods could be protected under trademark law. Those signs and symbols may either be biotechnology-products oriented or specialized ones.

Patent Protection for Biotechnological Innovations

The most widely recognized option for protection of biotechnological innovations is patents. Patent protection is generally preferable to publication, especially for the following categories of inventions: core or platform technologies with multiple applications; improvements to core technologies where the core technology is not already protected by patent; and improvements to patented inventions where patent term is important. The debate over patents on life forms of biotechnology has emerged largely as a response to the booming commercialization of biotechnology inventions over the past 30 odd years.[30] The TRIPS agreement[31] categorically considers patents as the suitable option for the protection of biotechnology inventions. Invention of a core technology with multiple commercial applications generally provides the foundation for broad patent protection. For example, discovery, expression, and sequencing of a gene associated with a disease is usually sufficient to obtain a patent covering the isolated gene, the synthetically produced gene, degenerate versions of the gene, the gene as a component of a heterogonous DNA sequence, the gene as a component of a plasmid, a cell comprising the plasmid, and so on. The system of patents has generated heated debates throughout the world in the recent past. In fact, the patent system has developed amidst many controversies. The issue of patenting of life and the inventions of biotechnology raised several questions on the rationality of the patent system. Granting patents on living systems or on life was criticized, since it involves conversion of life from nature's gift into private property. Patent is a monopoly right

---

[30] Ricardo Melendez and Sanchez, *Trading in Genes*, p. 141.

[31] See Wuger and Cottier. 2008. *Genetic Engineering and the World Trade System*, p. 78.

granted to the inventor for the contribution to the existing knowledge of the society in the form of a new invention. Patents are granted on the basis of certain broad principles that form the basics of patent law. The international law tries to ensure patent protection to novel innovations and technology. The TRIPS agreement is quite clear in its language under Article 27 when it says that patents shall be made available in all the fields of technology.[32] The principles of patent law broadly state that the invention must be a patentable subject matter, though it has not been properly defined and there is no fixed boundary for its purview. As a consequence, it was interpreted depending upon the existing needs. In the light of rapid developments in science and technology, its scope has been widened and expanded on several occasions giving rise to dramatic changes and new trends in patent law. A current development in the patent law and the trend of life patenting has brought changes in its philosophy and literally altered the contours of patent law, and it is a matter for concern. Patentable subject matter[33] has always been at the realm of interpretation and decision-making in terms of granting or not granting patents. The emergence of modern science called biotechnology made it possible to invent new and innovative nonnatural living beings. Since the times of evolution of patent law until the twentieth century, living beings were not considered patentable subject matter. Such developments of biotechnology were not expected and no provision was made in the then existing patent laws. Earlier to the dawn of biotechnology era, patenting of living beings was neither considered nor debated nor offered protection. However, it was the judiciary that took initiative to interpret the existing patent law in an innovative and liberal way, to offer patent protection to biotechnology inventions while extending the purview of patentable subject matter. In fact, more than legislative initiatives, judiciary's pronouncements have laid foundation for the evolution of biotechnology patent law. The judiciary did not alter or change any provisions of existing patent law; rather it has taken a liberal approach in interpreting the patentable subject matter and the patent law, which accommodated biotechnology innovations under the realm of patents.

---

[32] See Ricardo Melendez and Sanchez. 2007. *Trading in Genes*, p. 124.

[33] See Somsen, Han. 2007. *The Regulatory Challenge of Biotechnology: Human Genetics, Food and Patents*, p. 239. Edward Elgar, UK.

Patent Protection of Biotechnology Innovations in the US

Till the 1960s, the US did not witness any major development regarding the patenting of biotechnology inventions. The US judiciary, which is known for its innovative approach, did not come across any instance claiming living beings for patents. The US Patent Office reportedly granted two patents on living microorganisms in 1967 and 1968. The US Patent Law states: "whoever invents or discovers any new and useful process, machine, manufacture, or composition of matter, or any new and useful improvement thereof, may obtain a patent."[34] As a matter of practice, inventions relating to life were kept outside the purview of patentable subject matter. As per the strict interpretation of patent laws, living beings such as microorganisms, plants, and animals produced through natural or essentially biological processes are not patentable. However, later developments in the patent system in the US articulate that natural living beings are prohibited from patenting but nonnatural living beings produced through nonbiological processes such as biotechnological processes are patentable. This innovative approach has laid the foundation for the patenting of biotechnology products such as genetically modified organisms, which are nonnatural.[35] The beginning and evolution of biotechnology patent law can be traced to the US.[36] The US Constitution states: "the Congress shall have the power to promote the progress of science and useful arts, by securing for limited time to authors and inventors the exclusive right to their respective writings and discoveries."[37] The plain meaning of the aforementioned provision does not say anything about patenting of biotechnology. It merely signifies the Congress's power to promote the progress of science and arts by granting exclusive rights to the inventors for a limited time. The promotion of science is the threshold of patent law. The US Patent Act[38] states: whoever

[34] Sreenivasulu, N.S. and C. Basavaraju. 2005. Biotechnology and Patent Law, *World Journal of Intellectual Property Rights*, Jan–December: 1–2.

[35] Sreenivasulu, N.S. 2006. 'Biotechnology and Patent Protection', in C.B. Raju (ed.), *Intellectual Property Rights*, 1st edn. New Delhi: Serials Publications.

[36] Sahalia, Manu Luv. 2003. *Perspectives Intellectual Property Law, Many Sides to a Coin*, p. 171. Delhi: Universal Law Publishing Co. Pvt. Ltd.

[37] US Constitution, Article 1, Section 8.

[38] US Patent Law as modified in August 2005.

invents or discovers any new and useful process, machine, manufacture, or composition of matter or any new and useful improvement thereof, may obtain a patent, therefore, subject to the condition and requirements under the patent law.[39] Earlier in the nineteenth century, there were moves to patent living processes in the US. For the first time, the US Patent Office granted a patent to Louis Pasteur on "yeast" that is free from organic germs of disease.[40] It was the first incidence in the history of patent law where a living matter was granted patent. The patent office considered the invention as an article of manufacture within the meaning of patentable subject matter under the US patent law, but the fact that the invention was a living process was not taken into account. Patent was indeed granted on a living matter but the intention and approach was not in the favor of encouraging patents on living matter or living processes. Subsequently in *Exparte Latimer*[41] a patent claim for fiber found in the needle of *Pinus australis* was rejected. A general principle was set that plants were natural products and not subjected to patent protection.[42] The view taken was that patent on such fibers would lead to patents on the trees and plants of the earth, which is unreasonable. Besides, it was thought to be impossible to describe plants in a written form, as new plants may differ from old ones in color and perfume, which cannot be described in a written form. Hence, it was decided that plants as patent subject matter could not satisfy written description requirement under Section 112 of the US patent statute. The case law has postulated the following two principles.[43]

1.   Plants are products of nature, though artificially bred, and are not patentable.
2.   Plants were thought not amenable to the "written description requirement" of the patent law.

[39]   Section 101 of the US Patent Law.
[40]   US Patent No. 141,072 granted in the year 1873.
[41]   1889, Dec. Com. pat.123.
[42]   Thorne, Relation of patent law to natural products, 6j.pat off. Soc. 23,24 (1923).
[43]   See generally Chisum, Donald S., Craig Allen Nard, Herbert F. Schwartz, Pauline Newman, and Hieff F. Scott. 1998. *Principles of Patent Law-Cases and Materials*, p. 776. University Case Book Series, New York Foundation Press.

The aforementioned two principles were found to be hurdles for granting patent protection for plants.[44] Till 1930, living organisms were not considered as patentable. However, Plant Patents Act, 1930, extended patent protection to certain asexually reproduced plants, such as reproduction of plants by grafting and budding. The act was viewed as reflecting the legislative attention to the problems of patenting living organisms. It has also relaxed the requirement of written description by stating that reasonable written description is enough to grant patents.[45] The act's purpose was to solve the technical problem of description of plants as referred in the *Latimer*'s decision. In 1970, the Congress enacted the Plant Variety Protection Act, which again extended protection to certain new plant varieties capable of sexual reproduction, thus intending to promote the agricultural industry by encouraging patents on sexually reproduced plants. The 1930 Act as well as the 1970 Act did not show any indication of including bacteria within the purview of the patentable subject matter. The inclusion of bacteria within the purview of patent protection was considered at the time of passing of the 1970 Act but it did not get materialized. It seems the Congress had intended to include bacteria within the scope of patent protection at the time of passing the 1970 Act. Soon after the passing of the Plant Patent Act, 1930, there came an opportunity for the courts to interpret the patentable subject matter under Section 101 of the patent statute.[46] According to the patent statute, patentable subject matter encompasses any "new and useful process, machine, manufacture or composition of matter or any new and useful improvement thereof."

The term "manufacture" was interpreted in *American Fruit Growers Inc. v. Brogdex Co.*[47] in which patent was claimed for "coated oranges." The debate was whether the claimed invention came within the meaning of the term "manufacture." The court interpreted the term "manufacture" to mean the products or articles for use from raw or prepared materials by giving to these materials new forms, qualities, properties, or combinations whether by hand labor or by machinery. In the present case, the court rejected the claims on the ground that coated oranges were

[44] H.R. Rep. No. 1129, 71st cong., 2d Sess, 7–9 (1930).

[45] 35 U.S.C. Section 162.

[46] Sreenivasulu. 2008. *Biotechnology and Patent Law*, p. 21.

[47] 283 U.S 1.11.51 S.ct, 328, 330, 75 L.Ed.801 (1931).

not sufficiently modified, and oranges were not attributed considerable new qualities or properties or a combination of the two in order to be considered as a manufacture. The decision of the court infers that the inventors could have been granted the patent if their oranges were considerably modified to possess new qualities, properties, or a combination of the two. It seems that in the early part of the twentieth century itself, the US was inching toward patenting of living beings. In biotechnology inventions, existing living organisms or biological resources are used as raw materials, giving them a new look or features, which would generally be manipulation and manifestation of the existing living being. Such being the case, it can be inferred that biotechnological inventions do come within the purview of the term "manufacture", forming part of the patentable subject matter.

The Product of Nature Doctrine

The basic philosophy of conventional patent laws was that the life and living beings are products of nature, which are created by God. There is a doctrine under the conventional patent laws that states: biological material such as "microorganisms, plants, and animals" are products of nature and are not result of a creative process.[48] The product of nature doctrine was successfully used by courts and patent offices to deny patents on life and living beings. In *Funk Brothers Seeds Co. v. Kalo Inoculant Co.*,[49] the claim was on a mixed culture of different strains of microorganisms, each of which was useful to inoculate the roots of different species of leguminous plants, assisting the plants in nitrogen fixation. Different species of root-nodule bacteria existed in the nature. The applicants made efforts to combine those different species of bacteria in a mixed culture suitable for inoculating a range of crops, but their attempts failed because different species inhibited each other's effectiveness in combination.[50] The US Supreme Court refused to consider the claimed

[48] Somsen. 2007. *Regulatory Challenge of Biotechnology*, p. 239.

[49] See for more details: Ramakrishna, T. 2003. *Biotechnology and Intellectual Property Rights*, 1st edn, p. 22. Bengaluru: Center for Intellectual Property Rights and Advocacy, National Law School of India University.

[50] Sreenivasulu and Basavaraju. 2005. 'Biotechnology and Patent Law', pp. 1–2.

living beings as patentable subject matter and denied the patent on the ground that the product belonged to nature and there is no effort of the patentee to change its status of product of nature. At the time of the aforementioned decision, the concept of patenting of life was not known. However, a change was witnessed in the premises of patent laws later. Patent offices and law courts started reconsidering their stands on the patenting of life. In the late 1970s, a remarkable change was noted in the approach toward patenting life, in particular by the US judiciary, which laid the foundation for patenting of life. Meanwhile, in *Shell Development Co. v. Watson*[51] the term "composition of matter" under the patentable subject matter was interpreted to mean "all compositions of two or more substances and all composite articles, whether they be the results of chemical union or of mechanical mixture or whether they be gases, fluids powder or solids." Biotechnological inventions involve isolating genetic material from one species and incorporating it into an intended species or isolation and purification of genetic material from different regions of human or animal body. In this sense, biotechnology inventions are composition of genetic material from different species or different regions of the body of a living being. Hence, it can be inferred that the court signaled liberalization of patent laws to encourage new techniques and technologies. Following the suit, the Congress enacted the Patent Act, 1952, by codifying the patent laws. As an extension of the philosophy "ingenuity should receive liberal encouragement", the act amended the contents of patentable subject matter, thereby replacing the term "art" with "process." Seemingly, it was a leap forward to include biotechnological processes and inventions within the purview of patentable subject matter. Based on the presumption under the product of nature doctrine, many of the biotechnology patents successfully granted, such as the ones for "insulin, erythropoietin, and alpha interferon", were earlier contested on the grounds that they lack novelty and were obvious extensions of prior discoveries[52] or products of nature.[53]

---

[51] 149 F. Supp. 279, 280 (D.C. 1957) (citing 1 A. Deller, Walker on Patents, p. 55 (1st edn 1937).

[52] Kornberg, Arthur. 1995. *The Golden Helix: Inside the Biotech Ventures*, p. 233. CA, USA: University Science Books.

[53] Somsen. 2007. *Regulatory Challenge of Biotechnology*, p. 239.

The Product of Man Doctrine

As far as the evolution of biotechnology patent law is concerned, judicial pronouncements played a significant role than the legislature's initiative. As the lawmakers did not foresee the growth of biotechnology, there were no provisions in the patent laws passed decades ago to offer protection to biotechnology innovations. The legislature was not dynamic enough to react quickly to the challenges posed by biotechnology. In the absence of legal norms on the regulation of biotechnological inventions and in the light of the growing importance of the biotechnology industry, the judiciary shouldered the responsibility of recognizing and protecting biotechnology inventions. In fact, judiciary only interpreted the existing legislations liberally in an innovative way, and even laid down, still relevant, different principles and guidelines. The US judiciary was innovative enough to interpret the existing patent law in the light of constitutional mandates of promoting the progress of technology, useful arts, and science for the progress of the society. The judiciary laid down different principles and guidelines, which are still relevant. Judiciary with its innovative approach not only guaranteed protection for biotechnology inventions but it also followed constitution pursuit of the promotion of useful arts and science. It is because of the innovative and liberal approach that biotechnology inventions are protected under the patent laws till date. The beginning of the 1980s witnessed a complete departure from conventional patent laws to patent life. The product of nature doctrine opposes monopolization of life in the hands of private individuals, forming the basic premise in denying patents on life which was overruled by the US Supreme Court to grant patents on life. In fact, the *Red dove* case[54] raised few important questions: whether life forms a patentable subject matter? Whether life is patentable? The invention claimed for patent was a method of breeding of doves. Interestingly, the patent was rejected not on the ground of product of nature but on the ground that the breeding method was not repeatable. The patent office did not answer the questions and did not even say that the claimed living being were not patentable. It seemed the patent office deliberately

---

[54] Watal, Jayashree. 2002. *Intellectual Property Rights in the WTO and Developing Countries*, p. 152. New Delhi: Oxford University Press.

left the questions unanswered with a foresight of patenting biotechnology products such as genetically modified life forms in the near future. Although the patents were not granted on the claimed living being, the decision of the court kept the question of "patenting of life" alive.

Patenting Microorganisms

The TRIPS agreement includes patenting of microorganisms even when it provides flexibilities to the member states to exclude plants and animals from patentability.[55] Broadly speaking, life on the earth encompasses different living beings including plants, animals, and microorganisms such as fungi, algae, and bacteria. Life on the earth has been divided into two biological forms: lower life forms and higher life forms. All the single-cell organisms (microorganisms) such as bacteria, fungi, and algae, and certain multicellular organisms such as plants are considered as lower life forms. Apart from plants, other multicellular organisms such as animals and human beings are considered as higher life forms. The ability of animals and human beings to think and express differentiates them from plants, which are otherwise equally multicellular organisms. Strictly speaking, till the 1980s, life in any form was not considered as patentable. It was reported that in the early 1970s the German Federal Supreme Court would have upheld patent protection for new microorganisms, if the inventor could demonstrate a reproducible way for its generation. As the judiciary started moving toward considering life as patentable and the patent system was evolving to patent life in *Rank-Hovis McDougall Ltd*[56] in Australia, it was stated that a naturally occurring microorganism is not patentable, but a claim to a pure culture of such an organism in the presence of some specified ingredients would be patentable as would a new process that uses a naturally occurring microorganism to produce a new product. The decision implied that nonnatural and human-made living beings such as genetically engineered or modified microorganisms are patentable. This

---

[55] Wuger and Cottier. 2008. *Genetic Engineering and the World Trade System*, p. 78.

[56] See Loughtan, Patricia. 1998. *Intellectual Property-Creative and Marketing Rights*. Sydney, Australia: LBC Information Services.

approach of the judiciary, which gave rise to conflicting arguments, was intensively debated all over the world. In fact, the core issue of conflict was not the fulfillment of technical requirements of patentability, but the ethical issues involved in making life a patentable subject matter. In the 1970s, moves toward patenting living matter were witnessed, which laid foundation for the evolution of biotechnology patent law, as there were claims for patents on living matter. The debate on patenting life intensified with the decision of the US Supreme Court in the *Chakraborty* case[57] in which the inventor Chakraborty sought a patent on "genetically modified bacteria being capable of eating oil spills." The inventor claimed both the process of producing the bacteria and the bacteria. For the first time in the history of patent law, the question of living matter being patentable or not was answered in probable affirmation. The patent office allowed process claims for the method of producing bacteria but the claim for bacteria was rejected.[58] Patent was opposed on the ground of product of nature: that the invention is related to life and the life does not constitute patentable subject matter. An appeal was made to the Patent Office Board of Appeals against the decision of the patent office. However, the Board of Appeals also gave a decision against the inventor while affirming the rejection made by the patent office. The board viewed that patentable subject matter, as enshrined under Section 101 of the US Patent Code, does not encompass living things such as microorganisms. The board did not go into detail of whether the invention involves any inventive step or not and plainly rejected the claim. An appeal was made to the Supreme Court. There was much anxiety and the whole world was eagerly waiting for the decision. The court was confronted with few questions: is life patentable? Does life constitute patentable subject matter? By moving away from the conventional practices of patent law, the US Supreme Court held that nonnaturally occurring, man-made life forms such as biotechnologically produced and genetically engineered life forms (microorganisms) are patentable. The US Supreme Court, by 5:4 majority, overruled the patent examiners' decision to reject the

---

[57] Chisum, Donald S. 1998. *Principles of Patent Law.* New York: Foundation Press.

[58] Sreenivasulu and Basavaraju. 2005. 'Biotechnology and Patent Law', pp. 1–2.

patent and held that the US patent statute is broad enough to cover living organisms. The court viewed that constitution has granted wide powers to the Congress to promote the promotion and advancement of science. While determining whether the claimed invention comes within the meaning of patentable subject matter, the court in particular considered two terms, that is, "manufacture" and "composition of matter" under Section 101 of patent law. The court passed judgment in *American Fruit Growers*[59] while interpreting the term "manufacture" and in *Shell development*[60] while interpreting the term "composition of matter." The court interpreted the phrase "composition of matter" to encompass living matter produced through biotechnology. The court observed that the claimed bacterium is a composite mixture of the features of known species, put together into one species. It was viewed that Chakraborty had genetically modified the natural bacterium to possess the capacity to eat crude oil spills.[61] Upholding the philosophy of Thomas Jefferson, the author of the first Patent Act in the US, that "ingenuity should receive liberal encouragement," it was held that the invention was a result of human ingenuity and intellectual labor, which should receive liberal encouragement. The court accepted inventor's claim that the bacterium was a nonnaturally produced human made microorganism, not existing in the nature. Responding to the contentions that living matter is a product of nature and cannot be patented, the court held that Chakraborty's invention was not a product of nature; it was a product of man, and human role involved in the invention differentiated it from a product of nature. The court accepted that the starting point of the invention was a product of nature, but the inventor had added his ingenuity in making the bacterium possess the capacity to eat up oil spills with accuracy and pace. The bacterium was a product of nature till there was human intervention, after which it became a product of man.[62] The

[59] 283 U.S 1.11.51 S.Ct, 328, 330, 75 L.Ed.801 (1931).

[60] 149 F. Supp.279, 280 (D.C. 1957) (citing 1 A. Deller, Walker on Patents, p. 55, 1st edn 1937).

[61] In fact natural bacterium can also eat up oil spills, but the process is slow. Chakraborty's genetically modified bacterium was much faster in eating the oil spills.

[62] Rejecting the argument that microorganisms cannot qualify as patentable subject matter until the congress expressly authorizes such protection, the

US Patent Law[63] states that "any invention or discoveries any new and useful process, machine, manufacture or composition of matter or any new and useful improvement thereof, forms a patentable subject matter." The same provision was liberally interpreted to encompass nonnatural, man-made living beings like genetically engineered microorganisms as patentable. In response to the issue of product of nature, the court held that the microorganism in its claimed form does not exist in the nature; it is purely a product of human ingenuity that can be patentable even if it encompasses a living being. The court eventually overruled the long-standing premise under the conventional patent law that life is not patentable. The biotechnology industry, which is responsible for the innovations of nonnatural, human-made genetically engineered living beings, received a major boost from the decision. At the time of judgment, there were at least 100 patent applications pending before the US Patent and Trademarks Office (USPTO), which were later disposed on the rationality of the decision. The case had set a new trend and it steered the beginning of a new era of patent law. The decision has raised the curtain for the evolution of biotechnology patent law and opened the floodgates for biotechnology patents. The result is the plethora of patent applications before the patent offices throughout the world claiming genetically engineered microorganisms such as bacteria, algae, fungi, and so forth.

Patenting Genetically Modified Plants

The debate over patenting of living matter initiated by the *Chakraborty* decision encouraged the filing of patent applications claiming plants.

Court held that biotechnology was unforeseen by the congress when it enacted Section 101 patentable subject matter. However, the congress employed broad general language in drafting Section 101 precisely because such inventions are often unforeseeable. The court highlighted the constitution objective of promoting the progress of science and also the philosophy of Jefferson that ingenuity should receive liberal encouragement in interpreting the patentable subject matter to encompass living matter.

[63] US Patent Code 1952, Section 100.

The TRIPS agreement also contains optional provisions for plants.[64] *Chakraborty* impacted not only the interpretation of patentable subject matter under Section 101 of the US patent statute to encompass microorganisms but also furthered the debate over patenting of other living material including plants and animals. With the encouragement and inspiration from the *Chakraborty* decision, the scientific community, with extended enthusiasm and spirit, furthered their research in the field of biotechnology, extending its scope to cover other living beings.[65] The streamlined research in the field of agricultural biotechnology resulted in the innovation of transgenic plants in a hitherto unknown manner by combining the qualities of different plants and putting them together into an intended plant. The resulted plants are genetically modified to possess certain special features such as high yield, resistance to pests, and capacity to withstand drought conditions, which were not otherwise possessed by those plants. As the progress in biotechnology witnessed tremendous growth in the form of transgenic plants, claims for patenting of transgenic plants and related inventions started pouring in. The journey of patent law in extending its canopy to include life within its realm of patentable subject matter, had reached another landmark in considering nonnatural, human-made plants as patentable. It was for the first time in 1986 in *Exparte Hibberd*,[66] maize mutants, that is, a plant was claimed for a patent. The patent examiner rejected the patent application plainly on the ground that the claimed invention was a product of nature and therefore, not patentable. Furthermore, it was contended that there was another law, the US Plant Variety Rights Law, under which protection could be sought rather than seeking a patent. However, the US Patent Board of Appeals overturned the refusal of the examiner to grant patent on maize mutants, as it held that the claimed plant did not constitute product of nature, since it was nonnatural and human made. Eventually, the board laid foundation for patenting of nonnatural plants such as genetically engineered or biotechnologically produced plants. The board relying on the principle outlined in *Chakraborty* case that

---

[64] Wuger and Cottier. 2008. *Genetic Engineering and the World Trade System*, p. 78.

[65] See Sreenivasulu. 2011. *Intellectual Property Rights*.

[66] Exparte Hibberd 227 U.S.P.Q. 443 (Bd. Pat. App. 1985).

human ingenuity when added to a product of nature it becomes a man-made product, decided the claimed maize mutants as eligible for patent protection. The Board followed the suit in *Chakraborty* and extended the philosophy of *Chakraborty* to human-made transgenic plants.[67] Soon after this decision, the US Patent Office started entertaining patent applications claiming plants and related inventions.[68] An increase was seen in the patent applications claiming transgenic plants and plant cells and tissues. In *HiBred International* v. *JEM AG Supply, Inc., Supp*,[69] the US Court of Appeals for the federal circuit upheld patents on plants.[70] Therefore, now plants and related inventions such as plant cells and tissues are eligible for patenting. Biotechnology is versatile and dynamic, and has kept its pace in evolving. The progress of biotechnology made the law to run behind it for regulation. *Monsanto Canada Inc.* v. *Schmeiser*[71] laid down the approach of Canada, which often followed the US policy, toward GM crop. Monsanto marketed seeds of different GM crops, namely, "canola." The seeds found in the field of the respondent had the same genetic structure as of Monsanto seeds.[72] Monsanto sued Schmeiser for violating its patent rights on GM seeds of "canola." The court ruled in favor of the patent holder, making the respondent liable for growing the GM crops through the patented GM seeds "canola." Both the US and Canada seemed to have been following the policy of full recognition of patent rights on GMO and biotechnology, in particular agricultural biotechnology and transgenic plants.

Patenting Genetically Modified Animal

Animals are considered as higher life forms as they possess the capacity to think and express feelings unlike lower life forms such as plants and microorganisms. Once transgenic plants were granted patents,

---

[67]   See generally, Jayashree. 2002. *Intellectual Property Rights.*

[68]   With this required boost from the patent office, biotechnologists carried forward their research in the field, resulting in many transgenic plants and related inventions.

[69]   2d 794 (D.Iowa 1999) No.99-1035.

[70]   *J.E.M. Ag Supply, Inc.* v. *Pioneer Hi-Bred International, Inc.* 534 U.S. 124 (2001).

[71]   (CA) 2003, 2 FC 165.

[72]   Somsen. 2007. *Regulatory Challenge of Biotechnology*, p. 151.

biotechnologists started trying their hand in producing transgenic animals. The result was new forms of animals with foreign genes called genetically manipulated animals. The voyage of biotechnology patent law furthered to cover animals and related inventions. In fact the TRIPS agreement contained optional provisions that could cover GM animals.[73] The changes brought by the *Chakraborty* decision took another dimension when nonnatural animals were claimed for patent. In 1987, an animal was claimed for a patent, thereby igniting debate over the patenting of animals for the first time in the history of patent law. In *Exparte Allen*,[74] the claim was for a genetically modified oyster. The patent was opposed on the ground of product of nature. Opponents contended that an animal, being a creation of God, is a product of nature and cannot be patented. Meanwhile, the patent office rejected the claim for the invention on the ground of obviousness, that is, lack of inventive step. On appeal, the Board of Patent Appeals and Interferences confirmed the rejection. The question here was whether an animal constituted a patentable subject matter. Although the patent was refused, interestingly the board did not deny the patentability of nonnatural animal. The board rejected to affirm that the oysters were naturally occurring subject matter and consequently left the issue of patenting of living organisms unanswered.[75] In other sense, the board accepted that the claimed oyster was a nonnaturally occurring living matter, adding fuel to the debate on animal patenting. Nevertheless, progress in science and research has brought many nonnatural animals under patent. There was vehement pressure from the biotechnology industry to grant patents on animals in the light of enormous potential and utility of the claimed inventions. In the light of this development, the US Patent Office issued a statement saying that the Patent Office now considered nonnaturally produced non-human multicellular living organisms, including animals, as patentable. Reflecting the policy of the US Patent Office for the first time in the *Harward Oncomouse* case[76] a nonnatural animal was patented. The

[73] Wuger and Cottier. 2008. *Genetic Engineering and the World Trade System*, p. 78.

[74] 1987 USPO 2d 1425.

[75] See generally Sahalia. 2003. *Perspectives Intellectual Property Law*, p. 173.

[76] Cannon, Brian C. 1994. *Toward a Clear Standard of Obviousness for Biotechnology Patents*. Cornell Law Review, Cornell University.

claim was for a genetically modified mouse susceptible to cancer disease that was useful in cancer research and drug development. A foreign gene, that is, oncogene was incorporated into the mouse by genetically engineering it, which made it susceptible to cancer making the mouse useful in testing cancer-fighting drugs and technology. The patent was granted on the ground that the invention is a nonnaturally occurring and human-made living matter. The decision brought animals within the purview of patentable subject matter. There was a vehement opposition against granting patents on animals, questioning the rationality and ethics involved. In fact, there were more ethical objections on patenting of animals than on patenting of plants. Nevertheless, the US Patent Office granted patent on "Onco mouse," reflecting its change of approach and it opened the door for patenting of various nonnatural animals. As a result, many nonnaturally produced and genetically modified animals were patented. Till now, more than 16 patents were granted on inventions relating to transgenic mice alone. The race continued and patents were also granted on transgenic pig, rabbit, sheep, and many more. The voyage of patent law witnessed patents on transgenic sheep producing protein in its milk, transgenic pigs that produced low-cholesterol meat, and animals that produced bio-pharmaceuticals in bioreactors.[77] The race continued and patents were also granted on goat and cattle expressing diverse protein. The voyage of biotechnology patent law reached a stage wherein it offered patent protection to biotechnologically produced and genetically modified or transgenic animals. As per the observations, it was now possible to build a form with transgenic animals to which patent protection was offered.[78]

Patenting of Genetic Material and Biotechnology Processes

In developed countries such as the US, the biotechnology industry has grown as a major industry. Private and public entities have identified the significance of biotechnology in the day-to-day life and made huge investments in R&D. In the US, there is a major upsurge in the research

[77] Jayashree. 2002. *Intellectual Property Rights*, p. 164.

[78] Ganguli, Prabudha. 2001. *Intellectual Property Rights: Unleashing the Knowledge Economy*, 1st edn, pp. 125–6. New Delhi: Tata McGraw Hill.

on human genetic material, resulting in a number of patents being granted in this area. Every day many genes, DNA, and other human genetic material are being patented. A plethora of patent applications are pending before the US Patent Office claiming different biotechnology inventions, majority being claims on human genetic material. It is believed that in the year 2001 alone the US Patent Office awarded 20,000 gene patents and another 25,000 were pending at the same time.[79] Further, with the Herculean and daunting task of human genome project, it is believed that the decoding of entire human genome revolutionized medical science and individualized cures for many hereditary diseases are in the pipeline. Reportedly, more than 100,000 provisional patent applications on various segments of human genome are awaiting decision at the US Patent Office. Different types of inventions including transgenic microorganisms, plants, animals, and human genetic material are being claimed for patents. Protection is provided to different transgenic living beings, except transgenic humans, through patents. Today, various biotechnology innovations ranging from genetically modified plants and animals that can be seen through the naked eyes to microorganisms, such as bacteria, living cells, and genetic material such as gene, gene sequences, DNA, DNA sequences, and also fragments of DNA and genes that can only be seen with the help of microscopes are being patented. Furthermore, processes for isolating and employing genetic material for intended purposes, for producing genetically modifying living beings, and for treating genetic and hereditary diseases are being patented. The evolution of patent law on biotechnology started with patenting microorganisms and it has reached a stage where only a biotechnologically produced human being is left out from the purview of patenting. The TRIPS agreement provides patent protection for microorganisms, plants, and animals saying that patents shall be granted for all kinds of inventions in all fields of technology. The contribution of the US judiciary in the evolution of biotechnology patent law is being internationally recognized and has influenced the TRIPS agreement. It can be said that the TRIPS agreement has given international recognition

[79] Service, R.F. 2001. 'Proteomics: Gene and Protein Patents Get Ready to Go Head to Head', *Science*, 294(5549): 2082–3. See also http://www.ncbi.nlm. nih.gov/books/NBK19858 (last accessed 23 February 2016).

to the biotechnology patent law, which has evolved in the US. On the other hand, in Europe, transgenic cell lines developed from animal and human embryos were claimed for patent in *Edinburg Patent* case.[80] It was observed that such animal transgenic stem cells are allowed under the EU biotechnology directive, as the prohibition is only for human embryonic stem cells. On the other hand, there were other arguments that such prohibition should be viewed broadly to also include animal transgenic stem cells. It was a case where, for the first time, transgenic animal cells were granted patent with a disclaimer to human or animal embryonic stem cells.

Patenting Human Genetic Material

The race for patenting life continued and reached a stage where patents were being claimed on human genetic material.[81] For the first time in the history of patent law, a patent was granted on a human cell line in 1984. The cell line was found to be useful in producing a cancer-fighting protein, which was isolated and purified from the body of a patient named John Moore. In the light of vehement opposition centered on the ethics involved, the claimants contended that the claimed human cell line was not available in the nature in its isolated and purified form and involved efforts to isolate and purify genetic material The patent was granted on the claimed cell lines considering the benefit to the society as purified cell line claimed to be useful in fighting cancer. With this decision, human genetic material such as cells, genes, and DNA have become patentable. Following the decision of the patent office on granting patent on human cell lines, patents were granted on methods to isolate human genetic material and also on proteins produced by the human genetic material.[82] In the *In re o Farrell* case,[83] patent was granted on a method to produce a foreign protein in a transformed species of bacteria successfully. In

---

[80]   European Patent No. EPO 695351.'Isolation, Selection and Propagation of Animal Transgenic Stem Cells'.

[81]   Sreenivasulu and Basavaraju. 2005. 'Biotechnology and Patent Law'.

[82]   Cannon. 1994. *Toward a Clear Standard of Obviousness for Biotechnology Patents*.

[83]   In re Patrick H. O'farrell, Barry A. Polisky and David H. Gelfand 853 F. 2d 894 (Fed. Cir. 1988).

Amegan Inc. v. *Chugai Pharmaceuticals Co. Ltd*[84] patent was granted on a DNA sequence coding human protein erythropoietin[85] and on the protein itself, in a highly purified form. In another interesting development, the US Court of Federal Circuit in *In re Vaeck*[86] granted a patent on a chimeric gene, which is a combination of both an animal gene and a human gene. Again in *In re Bell*,[87] patent was granted on nucleic acid molecules DNA and RNA containing human sequences, proteins, and amino acid sequences coding for human insulin like growth factors, which play a role in the mediation of somatic cell growth following the administration of growth hormones. Further in *In re Deuel* patent was granted on DNA and cDNA encoding heparin-binding growth factors that stimulate cell division and facilitate the repair or replacement of damaged or diseased tissue. Meanwhile, in England in the *Novartis*[88] case human genes were held to constitute patentable subject matter. Further, in *Kingdom of the Netherlands* v. *European Parliament and Council of the European Union*,[89] while discussing on the status of the sequencing of human genes the court said that the result of such work could give rise to the grant of a patent only if the application was accompanied by both a description of the original method of sequencing that led to the invention and an explanation of its industrial application, as required by Article 5(3) of the EU directive. In the absence of an application in that form, there would be no invention but rather the discovery of a DNA sequence that would not be patentable as such. As a step further in *Pioneer Hi-bred International* v. *Holden Foundation Seeds Inc.*,[90] patent was claimed for human cloning methods.[91] However, at the same time it was held that human cloning

---

[84] *Amgen, Inc.* v. *Chugai Pharmaceutical Co., Ltd.* 808 F. Supp. 894 (D. Mass. 1992).

[85] Erythropoietin is a protein that boosts red blood cell production.

[86] In re Mark A. VAECK, Wipa Chungjatupornchai and Lee McIntosh. 947 F.2d 488 (1991).

[87] In re Rufus Bell et al. 991. F.2d 781 (Fed. Cir) 1993.

[88] *Novartis AG* v. *IVAX Pharmaceuticals UK Ltd* [2007] All ER (D) 252

[89] Case C-377/98, 1-07079.

[90] *Pioneer Hi-bred International* v. *Holden Foundation seeds Inc. 35 F.3d 1226, 31 U.S.P.Q.2d 1385, 39; Fed. R. Evid. Serv. 993* (hereinafter '*Pioneer Hi-bred International*').

[91] Jayashree. 2002. *Intellectual Property Rights*, pp. 151–2.

was prohibited except for therapeutic purposes. After the first isolation and culturing of human embryonic stem cells by Wisconsin biologist James Thomson in November 1998, embryonic stem cells have rapidly become one of the most exciting yet controversial areas of biomedical research.[92] As we understand, embryonic stem cells have the ability to develop into any one of the specialized cell or tissue types found within the human body and any research claims for patenting the same is considered as immoral. The same rationale was made even clearer in the *Wisconsin Alumni Research Foundation (WARF)* case,[93] in which patent application was on human embryonic stem cell cultures from primates, including humans. It was probably the first instance when human embryonic stem cells were claimed for patent in Europe. The application was rejected on the grounds of morality as human embryonic stem cells could not be derived without prior destruction of embryos.

Patenting of Biotechnology Processes and Methods

Along with human genetic material, methods or processes to isolate such material are also claimed for patents. Methods of expressing specific proteins in vitro (outside the animal or human body) through incorporation of gene coding are also claimed for patent. In order to produce a certain protein coded by a particular gene, the gene shall be isolated and incorporated into an intended body, may be, a bacterium. The bacterium produces the protein for which it was genetically engineered as a regular biological process. The methods to produce antibodies from the genetic material are also claimed for patent. In *Hybertech, Inc.* v. *Monoclonal Antibodies Inc.*,[94] the claim was for an immunoassay (method) that utilized monoclonal antibodies to measure the concentration of certain

---

[92]   See Plomer and Torremans. 2009. *Embryonic Stem Cell Patents, European Law and Ethics*, p. 22. London: Oxford University Press.

[93]   *Wisconsin Alumni Research Foundation (WARF)* v. *European Patent Office*, T 1374/04 (OJ EPO 2007, 313) (hereinafter '*Wisconsin Alumni Research Foundation*').

[94]   623 F. Supp. 1344 (1985). See also Chisum cases and materials on *Principles of Patent Law*, pp. 287–92. New York: New York Foundation Press, 1998.

antigens (viruses) in a given solution. Inventors produced a method to detect antigens by employing antibodies (proteins) produced inside the body and it also involved a method to produce antibodies to fight against antigens in a given solution. Patent was granted on the invention, which is first of its kind in developing a method to employ antibodies against antigens. It was for the first time that the patent office granted patent on a biotechnological method of employing antibodies produced by the body's immune system to fight against antigens. This patent resulted in a number of patent applications claiming different methods to raise antibodies and employ them against antigens. In *In re Wands*[95] the claim was for an "immunoassay" utilizing monoclonal high-affinity immunoglobulin antibodies. The invention involved immunoassay methods for the detection of hepatitis-B surface antigen by using high-affinity monoclonal antibodies. The invention was denied patent by the patent office on the ground of undue experimentation required to practice the invention. In an appeal made to Federal Circuit, it was viewed that the inventor had deposited the invention and the practice of the invention did require some experimentation. The invention was a method to produce proteins against specific viruses. The Federal Circuit, relying on the reasoning of the *Hybertech* decision, held that although some experimentation was required in practicing the invention, it was still patentable. This case was another example where a method to utilize antibodies against antigens was claimed. Perhaps the isolation and purification of genetic material is a Herculean task. In fact, genetic material is not available to be easily isolated. The intended genetic material must be identified before its different parts are isolated. For example, genes are not available to be isolated. A gene is not situated at a particular place. A gene comprises different fragments. Particular regions of DNA that express proteins are called genes. These genes are located in the form of fragments on different regions of a DNA. The isolation of a gene requires the isolation of its different fragments on DNA. After isolation, the material needs to be purified in order to incorporate it into

---

[95] In re Jack R. Wands, Vincent R. Zurawski, J.R., and Hubert J.P. Schoemaker 858 F.2d 731 (Fed. Cir. 1988). See generally Anderson, Debra Z. 2003. 'How to Enable Your Biotech Disclosure for United States Patent Application', *Lawyers Journal*, Allegheny County Bar Association.

an intended body for performing an intended task such as producing a specific protein. As per the latest decisions of the Court of Federal Circuit and the decisions of US Patent Office, biotechnology processes or methods for isolation and incorporation of genetic material from one species to another constitute a patentable subject matter.[96] In Europe in the *Wisconisin Alumni Research Foundation* (WARF) case[97] the patent application was on the method for obtaining embryonic stem cell cultures from primates, including humans. It was probably the first instance when methods for obtaining human embryonic stem cells were claimed for patent in Europe. The application was rejected on the grounds of morality, as such methods for obtaining human embryonic stem cells could not be practiced without prior destruction of embryos. The upsurge of biotechnology patents include patent granted to the Cetus Corporation on the technique of amplifying tiny fragments of DNA in 1985 and recombinant DNA patent to Stanford for the use of rDNA to clone genes in bacterial, plant, and animal cells.[98] Many countries such as Japan, Australia, New Zealand, Korea, and Singapore have laws similar to the US on the patenting of biotechnological inventions.[99]

Patent Law and Biotechnology in India

The Patent Act, 1970, highlights that an invention that meets the universally accepted requirements of patentability such as novelty, inventive step, and industrial application is patentable. The act defines invention to mean any new and useful art, process, method or manner of manufacture, machine, apparatus, or substances produced by manufacture and includes any new and useful improvement in any of them. The act does not specify the inventions that are patentable, but it illustrates subject matters that are not patentable. The act allows patenting of processes but it does not consider patenting of products. However, the Indian Patent Act enacted

---

[96] In *In re vaeck* the claim related to manipulating a bacterium to insert a gene. The inventors isolated a gene and combined it with a stretch of DNA from the bacteria and created chimeric gene. A patent was claimed in 1991 and the same was granted.

[97] 2006, IJ EPO 393.

[98] Kornberg. 1995. *The Golden Helix*, pp. 236, 241.

[99] Ricardo Melendez and Sanchez. 2007. *Trading in Genes*, p. 127.

in 1970 does not mention anything about biotechnological invention as biotechnology was not developed in India at the time. After biotechnology industry started flourishing in the US and the EU, it gained some momentum in India too. Once patents were granted for different biotechnology inventions in the US and the EU, the demand to adopt the same approach resonated throughout the world, including India. As a matter of fact, the judiciary was responsible for the evolution of patent law on biotechnology invention in the US and the EU. Accordingly, patent laws were amended and a few specific legislations were brought up to grant patents on biotechnology inventions. This momentum set new trends in the history of patent law and influenced the adoption of international conventions such as the TRIPS agreement. The agreement states that patents shall be made available to all types of inventions in all fields of science and technology. The agreement mandates patenting of biotechnology inventions in the member states.[100] After ratifying TRIPS, India modified all its intellectual property laws, including patent law, and started marching toward patenting of biotechnological inventions. To fulfill its obligation under the TRIPS agreement to provide patent protection to all kinds of invention in all the fields of science and technology including biotechnology inventions, India has amended its patent law thrice. The first amendment was made in 1999, though this amendment was brought into effect retrospectively from 1995. The amendment brought changes in the original act to extend patent protection to inventions relating to manufacture of chemical substances intended for use or capable of being used as food or as drug. Apart from all chemical substances that are ordinarily used as intermediates in the manufacture of any of the medicines or substances defined under Section 2(1) (1) of the 1970 Act, invention of a substance intended for use or capable of being used as medicine or drug could be patented.[101] In fact, the 1999 amendment provides only for accepting patent applications claiming for inventions related to agriculture and pharmaceuticals. After the amendment, thousands of applications were filed before the Indian patent office,[102] among

---

[100] Article 27 of the TRIPS agreement.

[101] See Section 5(2) of the Patents Act.

[102] In India, the main patent office is situated in Calcutta; however, there are regional offices at Delhi, Bombay, and Chennai. Applicants can file their applications at any of the aforementioned patent offices.

which a considerable number of applications were related to biotechnology inventions, such as chemical substances isolated from the body of living beings. However, all these applications were kept in a mailbox that was opened on 1 January 2005 and processed. Till the grant of patents, the applicants were given marketing rights on their inventions if they were patented in other countries. India again amended its patent law in 2002,[103] bringing major changes to allow patenting of living beings such as microorganisms, plants, and animals produced through biotechnological processes. Till the time of this amendment, the term "invention" under the Patent Act encompassed only processes but the amendment included the term "product" within the meaning of invention. After the amendment, the definition for invention stands as follows: invention means a new "product" or process involving an inventive step and capable of industrial application.[104] The amendment added the following provisions to Section 3 of the Act, which talk about excluded subject matters from the Indian patent law purview. It is significant in the light of patenting of biotechnology inventions.[105]

1.   an invention, the primary or intended use of commercial exploitation of which could be contrary to public order or morality or which causes serious prejudice to human, animal, or plant life or health or to the environment;

2.   discovery of any living thing or nonliving substance occurring in nature;

3.   therapeutic and diagnostic or other treatment of human beings or animals;

4.   plants and animals in whole or any part thereof other than "microorganisms," but including seeds, varieties, and species and essentially biological processes for production or propagation of plants and animals.

The amendment states that microorganisms are patentable but plant and animal varieties and essentially biological processes for

---

[103]   Patents Amendment Act, 2002.

[104]   The amendment includes the word "product" in the definition of invention to imply that patents are made available to products or processes. See the definition of invention under the Patents Act 1970.

[105]   Section 3 of the Patent Act as amended in 2002.

the production of the same are excluded from the purview of patent protection. However, it is mentioned that this exclusion does not apply to microbiological processes and products thereof. Technical interference to a natural or essentially biological process renders it a nonbiological process; products of biotechnological process produced through some technical interference to natural process are patentable. Chemical processes are patentable under the original act; however, the 2002 amendment added explanation to chemical process, which states: "Chemical processes include biochemical, biotechnological and microbiological processes."[106] That means, along with chemical processes, bio-chemical and microbiological processes are also patentable. In March 2005, India amended its patent law for the third time.[107] The amendment recognizes the Budapest Treaty[108] for the deposit of microorganisms for the purpose of patent procedure. As it may not be possible to describe a biotechnology invention as per the requirements of the patent law, the treaty recognizes the practice of depositing the invention in any depository recognized under the treaty to compensate the written description. In order to fulfill the TRIPS mandates of providing patent protection to biotechnology inventions, India recognized the Budapest Treaty, which facilitates for the deposit of biotechnology inventions. Furthermore, the 2005 amendment totally omits Section 5 of the original act, which restricted patents only to the methods or processes of manufacture. The 1999 amendment altered this section by inserting subclause (2), making both products as well as processes patentable in India.[109] Developing countries such as Argentina, Brazil, and the Andean Group have implemented the TRIPS agreement and allow patents only for microorganisms and microbiological processes, excluding plants, animals, genes, and other biological material, just like India.[110]

[106] Explanation to Section 5 of the Act as amended in 2002.

[107] Patents Amendment Act 2005, published in the Official Gazette of India on 5 April 2005.

[108] The Budapest Treaty on the International Recognition of the Deposit of Microorganisms for the Purposes of Patent Procedure was adopted in the year 1977; India is a party to this treaty.

[109] Patents Act as amended latest in March 2005.

[110] Ricardo Melendez and Sanchez. 2007. *Trading in Genes*, p. 127.

Protection of Plant Varieties and Farmers' Rights Act, 2001

The Plant Varieties and Farmers Rights Act deal primarily with the protection of plant breeders' rights over the new varieties developed by them and the entitlement of farmers to register new varieties and to save, breed, use, exchange, share, or sell the plant varieties, which the latter have developed, improved, and maintained over many generations. The act is a deviation from the 1991 International Union for the Protection of New Varieties of Plants (UPOV) model and can be regarded as an alternate sui generis system that accords protection to the rights of the formal innovations of a plant breeder and the informal knowledge system and traditional plant varieties of the farmers as well.[111] There are important provisions contained in this act on the protection of farmers' rights and the mechanism suggested for compensation or benefit-sharing for the contributions of local communities or farmers in the development of a new plant variety.[112] This act established an effective system for the development and protection of plant varieties, and the rights of farmers and plant breeders.[113] Such protection is likely to facilitate the growth of the seed industry, which will ensure the availability of high-quality seeds and planting materials to farmers.[114] The protection of the rights of the farmers is a significant step as it recognizes the place of the farmers in the production of seeds and their contribution in conserving plant genetic resources.[115] Breeders or farmers will get rights to produce, sell, market, distribute, import, or export the varieties. This legislation is a

[111] See Pushpangadan, P. and K.N. Nair. 2005. 'Value Addition and Commercialization of Biodiversity and Associated Traditional Knowledge in the Context of the Intellectual Property Regime', *Journal of Intellectual Property Rights*, 10: 441–53.

[112] Sreenivasulu, N.S. and Arnab Sengupata. 2010. 'Biological Resources, IPR and Biodiversity', *Manupatra Intellectual Property Reports*, 1(4): F-35.

[113] Sreenivasulu, N.S. and K.S. Kariyanna 2012. 'Biological Diversity, Intellectual Property and Patents: Concerns of Biological Resources', *Manupatra Intellectual Property Reports*, 1(2): F-47.

[114] Wadehra, B.L. 2010. *Intellectual Property Rights*, 4th edn, p. 499. New Delhi: Universal Law Publishing Co. Pvt. Ltd.

[115] Ramappa, T. 2010. *Intellectual Property Rights Law in India*, 1st edn, p. 311. Hyderabad: Asia Law House.

government response to the obligation created under Article 27.3b of the TRIPS agreement, wherein protection has been granted to the "creator" breeding the seeds through the Plant Breeder Rights (PBRs) in order to provide incentive to the private sector actors to engage in plant breeding and achieve the ultimate goal of enhanced food security through provision of new improved varieties and availability of seeds through the private sector channels.[116] In India, knowledge on the conservation and use of plant genetic resources has traditionally been community knowledge, without it being accrued to a particular entity, and hence the issue here is that of the acceptability of the entire concept of IPRs in agriculture and the need to establish PBR of the nature that the act allows for.[117] Since farmers traditionally reuse the seeds from their harvests, they are considered direct competitors of breeders who develop plant varieties for commercial interests and then seek legal protection for the exclusive market exploitation of their varieties, as such a regime takes away the traditional and community-centered control over seed conservation and use, which has been the regular practice of farming communities all over the country.[118] Although farmers are given rights under Section 39 of Protection of Plant Varieties and Farmers' Rights Act, 2001, to acquire protection similar to the plant breeders in case of a new variety bred and developed by the former, but such an option would not be feasible as only those farmers who are rich and seek to acquire a stake in the food production will avail of such a mechanism. The Farmers' Rights definition in the act adopted and expanded all three aspects:

1. Farmers' privilege as a right is not only to save and exchange seeds but also to sell seeds (except branded).

[116] Cullet, Philip and Radhika Koluru. 2003 [2002]. 'Plant Varieties Protection and Farmer's Rights: Towards a Broader Understanding', *Delhi Law Review*, 24: 41.

[117] Bhutani, Shalini and Kanchi Kohli. 2004. 'Protection of Plant Varieties and Farmers' Rights Act, 2001—Just Leave the Seed Alone', *The Hindu Business Line*, 12 March, available at http://www.thehindubusinessline.com/2004/03/12/stories/ 2004031200100900.htm (last accessed 23 February 2016).

[118] Bhutani and Kohli. 2004. 'Protection of Plant Varieties and Farmers' Rights Act, 2001—Just leave the seed alone.'

2. Benefit sharing based on compensation and operating through a mechanism where communities/farmers can make claims for such compensation.
3. Farmers' Rights to ownership, as the idea that farmers must be able to register their varieties.[119]

The protection thus granted to genetically modified seeds and plants is posing a threat to farmers' rights, as they are unable to reuse and resell the protected seeds and are forced to purchase from the open market each year.

For biotechnology innovations, patents are claimed for both processes as well as products. The amendments made to the patent law and the emergence of biotechnology patent law in India can be traced back to the TRIPS agreement. However, before implementing the TRIPS agreement, a patent on a living being was already granted. Dominico, a subsidiary of American Home Remedies, was granted a patent on a process for the preparation of infectious bursitis vaccine. The vaccine is useful in protecting poultry against infectious bursitis. Initially, the patent office rejected patent and an appeal was made to the High Court of Kolkata.[120] The high court held that there was no bar for patenting a claim, which was a living being or living process. The court interpreted the term "manufacture" under the patent law to include any living being or living process and directed the patent office to grant the patent. It was the first instance where a living process was patented in India. With the amendments made to the patent law fulfilling obligations under the TRIPS agreement, biotechnology patent law emerged in India. Furthermore, the Protection of Plant Varieties and Farmers' Rights Act provides for the sui generis mechanisms for the protection of plant varieties and plant biotechnology innovations. This legislation is perceived

[119] Ramanna, Anitha. 2003. 'India's Plant Variety and Farmer's Rights Legislation: Potential Impact on Stakeholder Access to Genetic Resources', EPTD Discussion Paper No. 96, Environment and Production Technology Division, International Food Policy Research Institute, Washington DC, available at http://ageconsearch.umn.edu/bitstream/16105/1/dp020096.pdf (last accessed 23 February 2016).

[120] Ramakrishna. 2003. *Biotechnology and Intellectual Property Rights*, 1st edn. Bengaluru: National Law School of India University.

to be the most customized legal framework meant for recognizing innovative efforts in the field of agriculture and plant biotechnology. Although it covers certain objectives of the UPOV convention on plant varieties and breeders' rights, it goes beyond the convention by providing protection for farmers' innovations and rights, which is an outstanding element in the legislation.

Presently, the Indian intellectual property law does not allow patenting of medical methods and agricultural methods.[121] Biotechnological patent in the field of agriculture is improbable under the present law due to the farming and agriculture sector needs of Indian society. New concepts and technology involved in this field take some time to gain public acceptance and adaption. Therefore, due to the slow acceptance rate, time will play a crucial role in finding the right balance. Much of the agitation is, in fact, due to the fear of letting multinational companies (MNCs) have the monopoly. There are many ethical issues involved as well, especially on humanistic grounds. Biotechnology as a tool has great potential for overcoming the constraints to increase agricultural production, but for this, substantial investments are required. India needs to ensure a clear regulation of IPRs in order to promote good private sector involvement. It is felt that having an intellectual property protection system will incentivize investment and research. Perhaps, contrary to the apprehensions, the current intellectual property system through compulsory licensing, and so forth, has enough means to ensure that monopoly is not exploited, and thus the lawmakers should strive to shift the current paradigm by granting intellectual property protection.

Patent Law and Biotechnology in Europe

The Paris Convention for the Protection of Industrial Property was adopted in 1883 and the Strasbourg convention[122] was adopted in 1963 with an objective of establishing a common market in Europe, whereas the European Patent Convention (EPC) of 1973 provide for the broader boundaries of patent law and common patent granting system in all the

---

[121] See Section 3(h), Indian Patents Act, 1970.

[122] Strasbourg convention on the unification of certain points of substantive law on patents for invention 1963.

European countries. The convention says that European patents shall be granted for any inventions that are new and involve an inventive step with industrial applications.[123] The convention prohibits patenting of living beings produced through essentially biological processes, but it does not prohibit living beings produced out of nonbiological processes.[124] In Europe, even before the adoption of the convention, patents were claimed on living matter produced through nonbiological processes. The convention does not state the inventions that are patentable; instead, it illustrates the inventions that are not patentable, for example, discoveries, scientific theories, and mathematical methods are the few mentioned in the list of inventions not patentable.[125] Furthermore, the convention states under Article 53(b) that: "Plants or animal varieties and essentially biological processes for the production of plants and animals are not patentable. However, this provision does not apply to microbiological processes or products thereof."[126] At the same time the convention states under Article 53(a) that: "inventions against public order and morality are not patentable under the convention."[127] The inventions falling under exception clause do not qualify for patent, even if they meet the requirements of novelty, industrial application, and inventive step. When we see the other side of the coin, it seems that "plants and animal varieties produced through non-essentially biological processes are patentable" under the convention. As only essentially biological processes and products thereof are not patentable, it can be inferred that all the other processes such as microbiological and biotechnological processes and products[128] thereof are patentable. On the same

---

[123]   European Patent Convention, Article 52(1).

[124]   European Patent Convention, Article 53.

[125]   European Patent Convention, Article 52.

[126]   European Patent Convention, Article 53(b).

[127]   European Patent Convention, Article 53(a).

[128]   Microbiological process is defined under the convention to mean "any process involving or performed upon or resulting in microbiological material." See Chapter VI, Rule 23B(6) of EPC, see also Article 2(1)(b). Biotechnology invention is defined under the European Patent Convention to mean "any invention which concern a product consisting of or containing biological material or process by means of which biological material is produced, processed or used." See Chapter VI, Rule 23B(2) of EPC.

lines, it can also be inferred that products of biotechnological processes or inventions of biotechnology such as transgenic microorganisms, transgenic plants, transgenic animals, and isolated and purified human genetic material are patentable. Even the European Patent Office (EPO) tried to interpret the convention liberally to patent inventions produced out of nonbiological processes. Later, the TRIPS agreement echoed the language of EPC regarding patenting of living beings. These three conventions played a vital role in streamlining the law on patents in Europe. In particular, EPC seems to have considered the issue of patenting of biotechnological processes and products.

Patenting of Microorganisms

In fact, Europe was ahead of the US in patenting living beings. The first landmark court ruling concerning patenting of life came about in 1969 in Germany.[129] In Germany, a patent was claimed on a method for breeding doves with red plumage.[130] The German Patent Office rejected the patent on the ground that the method was not repeatable and the Supreme Court of Germany confirmed the same. It was the first case where a process for producing living matter was claimed in the EU. Seemingly it was the first case to open the door to the patenting of biotechnology inventions.[131] Further, in the early 1970s, five years before the *Chakraborty* case, German Federal Supreme Court would have upheld patent protection for new microorganisms if the inventor would have been able to demonstrate a reproducible way for its generation.[132] Further, EPC adopted by the EU lays down a comprehensive framework on the law related to patents. The aforementioned decision of the German Supreme Court, that is the rejection of the claim, was proved controversial in the light of EPC, excluding the patenting of plant and animal varieties or essentially biological process for their production. The German Supreme Court's decision can be viewed from the positive side of the EPC. The convention

[129] Ricardo Melendez and Sanchez. 2007. *Trading in Genes*, p. 141.

[130] Adelman, Martin J., Randal R. Reader, John R. Thomas, and Harald C. Wagnar. 2003. *Cases and Materials on Patent Law*, American Case Book Series, 2nd edn, p. 156. Minnesota, USA: Thomson West.

[131] See generally, Jayashree. 2002. *Intellectual Property Rights*.

[132] Wagner. 1976, p. 335.

excludes patents for products of essentially biological processes but it does not exclude patenting of products of nonbiological processes. Therefore, it can be inferred indirectly that the language of EPC laid the foundation for patenting of living beings in Europe. As mentioned, since the TRIPS agreement adopted the language of EPC regarding patenting life including biotechnologically produced living beings, it can be said that patenting of biotechnology innovations derived first affirmation through the language of the EPC. Meanwhile, for the first time after the adoption of the EPC, which seemingly laid the foundation for the patenting of living beings, a living being was claimed for patent in *Genentech-I/ Polypeptide expression*.[133] The invention was a plasmid, a microorganism suitable for transforming a bacterial host, which included an expression control sequence or "regulon" that could enable the expression of foreign DNA as a recoverable polypeptide. The applicants invented a general principle for enabling plasmids to control the expression of polypeptides (proteins) in bacteria. The European Patent Appeals Board opined that only essentially biological processes and products thereof are excluded under the convention, but non-biological processes and products thereof are patentable under the convention. The board viewed the present invention as a product of biotechnological processes, which, in the opinion of the board, does not constitute essentially biological processes and hence is patentable. The decision of the board represents approach of the patent office in patenting products of processes other than essentially biological processes. The decision assured the inventors that patents were granted for products of biotechnological processes and established the rule that microorganisms were patentable in the EU.

Patenting of Biotechnological Processes

At the time when the EPO Board allowed a patent on a plasmid, the biotechnology industry got a setback in the UK. The Court of Appeal in the UK rejected a patent on various claims related to the production of a substance called human tissue plasminogen activator (tPA) in *Genentech Inc v. The Welcome Foundation*.[134] The invention was useful in treating

---

[133] *Genentech-I/Polypeptide expression v. European Patent Office*, [1989] O.J. EPO 275.
[134] (1989) R.P.C 147.

a disease thrombosis and hence was of great commercial value.[135] The human tPA was produced through rDNA technology. At the time of the invention, DNA technology existed and, at least in theory, could be applied to produce tPA as the same technology was used to produce growth hormones, insulin, and other valuable products of similar nature. Therefore, it was believed that there was no inventive step involved in the invention and the patent was rejected for the lack of inventive step. This decision was seen as an anti-biotechnology trend and it generated chaos in the industry.

However, the industry regained confidence with the decision of the same Court of Appeal in *Chiron Corporation v. Murex Diagnostics Ltd & Organon Teknika Ltd.*[136] The *Chiron* case held claims for producing vaccines, polypeptides, and antibodies through a biotechnology process to fight against hepatitis C virus. Chiron owned patents, which included claims to polypeptides, immunoassays, polynucleotides, antibodies, vaccines, and methods of in vitro propagation of the hepatitis C virus.[137] Chiron was successful in identifying and sequencing the virus responsible for causing non-A, non-B hepatitis (NANBH). The claims involved an immunoassay kit (method) for detecting antibodies against hepatitis C virus by screening blood for the presence of hepatitis C infection, a vaccine, and a method of cultivating hepatitis C cells. The invention was considered eligible for patent as it met the requirements of novelty, industrial application, and inventive step. It was held that the invention was hitherto unknown and novel; it was the first-of-its-kind invention of a method to detect antibodies against hepatitis virus to treat hepatitis C. The court acknowledging EPC held that the claimed biotechnological process was eligible for the grant of patent. The decision established a rule that inventions related to biotechnological processes were patentable.[138] In *Genentech*, the Court of Appeal was strict in

---

[135] See generally Ramakrishna, T. (ed.). 2003. *Biotechnology and Intellectual Property Rights*, 1st edn, pp. 28–9. Bengaluru: Center for Intellectual Property Rights and Advocacy (CIPRA), National Law School of India University.

[136] (1996) R.P.C. 535, CA.

[137] The patent was applied for in the year 1987 and granted in the year 1990.

[138] See generally, Hollyoak, Adam and John Torreman. *Intellectual Property Law*, 2nd edn, pp. 74–5. London: Butterworths.

applying existing patent law on biotechnological inventions. However, the EPO Board was liberal when it encouraged patents on biotechnology inventions in *Chiron Corporation*. In the absence of specificity and unsettled legal opinion on the patenting of biotechnology inventions, courts have adopted different views and opinions, and it is possible that approaches of two judges may run contrary to one another. Therefore, the approach of the judiciary may not remain the same as the judges may not hold their office for a long time. So, when a new judge enters the office, he or she may express a contrary view for the case of similar nature in the absence of specific legislation laying down standard law. *Genentech* and *Chiron* were two different strides in the evolution of biotechnology patent law in Europe reflecting the fluctuating views of the judiciary in the absence of specific legal norms on the issue. However, by looking at the approach of the EPO and the decision in the *Chiron* case, it can be inferred that biotechnological processes are patentable in the EU.

Patenting of Transgenic Plants

The progress of biotechnology industry in Europe resulted in research on the production of transgenic plants. As genetic manipulation in microorganisms was a success, the biotechnology industry genetically manipulated plants to invent new varieties of plants.[139] Encouragement was sought for research on transgenic plants through grant of patents. On the basis of the rationality in granting patents on microorganisms and biotechnological processes, patents were being claimed on plants produced through biotechnological processes. The EPO is in favor of applying the rationality of patenting microorganisms and biotechnological processes to the plants as it might stimulate further research in the industry on transgenic plants. In *Ciba Geigg*,[140] the EPO Technical Board of Appeal[141] granted a patent on a plant.[142] The invention was

[139] Sreenivasulu and Basavaraju. 2005. 'Biotechnology and Patent Law', pp. 1–2.

[140] *Ciba Geigg v. European Patent Office*, Case T 49/83 [1984] O.J. EPOs 112 (hereinafter 'Ciba Geigg').

[141] (1984) O.J EPO 112, Tech, Bd. App.

[142] See generally, Cornish, W.R. 1999. *W.R Cases and Materials on Intellectual Property*, 3rd edn, pp. 60–2. London: Sweet and Maxwell Limited.

propagating material for cultivated plants, in particular, seeds of cultivated plants treated with a chemical agent, that is, oxime derivative, in
order to make it resistant to agricultural chemicals. A definition of cultivated plants in the description includes all plants that yield substances
in any form. Propagating material from such cultivated plants comprises
all reproductive plant components, including plants and plantings that
have begun to be germinated, but particularly seeds. The subject matter
of claims is not an individual variety of plant distinguishable from any
other variety, but the claims relate to any cultivated plants in the form
of their propagating material, which have been chemically treated in a
certain way. By referring to the exclusions under the EPC, the board
observed that EPC Article 53(b) prohibits only the patenting of plants
or their propagating material in the genetically fixed form of the plant
variety. It further opined that the propagating material claimed was
not the result of an essentially biological process, but was a nonnatural
process of treating cultivated plants in the form of their propagating
material with a chemical according to a set formula. The board held that
the claimed invention is a nonnatural plant, which is a result of treatment with chemical agents. The board, by interpreting the Article 53(b)
of EPC restrictively, held that nonnatural plants do not fall within the
purview of exceptions and are patentable.

It was the first case where nonnatural plants produced through a
biotechnological process were claimed and patented in Europe. Further
in 1993, in Plant Genetic Systems[143] genetically modified plant cells and
seeds were claimed for patent. Patent protection was claimed for plants,
plant cells, and seeds possessing a foreign gene capable of resisting a
type of herbicide, namely, glutamine synthesis inhibitors (GSI).[144] The
claimed patent discloses the use of biotechnological techniques for the
production of GSI-resistant plants and seeds that contain heterologous
DNA encoding a protein capable of inactivating or neutralizing the
above-mentioned herbicides. A new trait is added to the genetic material

---

[143] Plant Genetic Systems/Glutamine Synthesis Inhibitors (1993) 24 IIC
618.
[144] However, such plants could also be obtained by traditional plant selection methods. Some plants may be naturally resistant, and some plants may
develop resistance.

of a plant, which allows the plant to grow in the presence of GSI. The claim of the invention was opposed by reason of exception under Article 53 of the EPC, which states that, invention, which is contrary to order public or morality is excluded from the purview of patent protection. The issues before the court for decision were of much importance. This case happens to discuss all the provisions in exception to patentability. In the light of ranging demand for patenting of living beings including plants, the decision of the case was very significant. The case has defined important terms under the EPC and settled many issues that were troubling the patent office in granting biotechnology patents. Irrespective of the outcome, this case has discussed in detail about what constitutes "order public" or "morality," "plant varieties,"[145] "essentially biological processes,"[146] and "microbiological processes."[147] Expressing its view on the concept "order public" the European Technical Board of Appeal held that the "order public" covers the protection of public security, physical integrity of individuals, and also the protection of the environment. In that sense, any invention whose exploitation is likely to breach public peace or social order threatening the public security or which may seriously prejudice the environment is to be excluded from patentability as being contrary to "order public."

Discussing the concept of "morality" it was held that morality is related to the belief that some behavior is right and acceptable, whereas some other behavior is wrong and not acceptable. This belief is founded on the totality of the accepted norms that are deeply rooted in a particular culture. Therefore, any invention that is not in conformity with the conventionally accepted standards of conduct pertaining to the culture is to be excluded from patentability. Deciding on the issue whether patents on seeds and plants go against public order and morality, it was

---

[145] Plant variety is defined under Article 5 of the Community Plant Variety Rights Regulation 2100/94. Plant variety is also defined under amendments made to EPC to adopt the provisions of the directive on the legal protection of biotechnological inventions.

[146] Essentially a biological process is defined under the recently enacted directive on the legal protection of biotechnological inventions.

[147] Microbiological process is defined under the recently enacted directive on the legal protection of biotechnological inventions to mean "any process involving or performed upon or resulting in microbiological material."

viewed that seeds and plants shall not constitute an exception to patentability under EPC Article 53(b) merely because they represent "living matter" on the ground that plant genetic resources should remain the "common heritage of mankind." A survey and opinion poll conducted in Sweden and Switzerland showed that farmers are neither willing nor against genetic engineering and production of super crops such as herbicide-resistant plants. At the same time, the survey results showed that farmers were against patenting of animals and plants. Considering the survey and opinion poll result, the board held that survey and opinion poll, being subject to influences and fluctuations within short-time periods in an unforeseen manner, do not necessarily reflect "order public" or moral norms and can scarcely be considered decisive. Expressing concern over genetically modified plants and biotechnology patents, the board opined that plant biotechnology cannot be regarded as being more contrary to morality than traditional selective breeding. It was noted that the objective of both traditional breeders and molecular biologists or biotechnologists was to change the property of a plant by introducing novel genetic material in order to obtain a new and possibly improved plant. It was also observed that comparatively genetic engineering techniques of biotechnology are more powerful and capable of giving accurate results and control over genetic modifications. While deciding whether plant cells and seeds do constitute "plant variety," which is excluded under EPC Article 53(b), the board held that any claim on or encompassing plant variety is not patentable. The board defined a plant variety to mean: any plant grouping within a single botanical taxon of the lowest known rank characterized by at least one single transmissible characteristic distinguishing it from other plant groupings and which is sufficiently homogeneous and stable.

However, the board held that genetically modified plant cells could not be considered plant varieties. In the light of the current practice of the EPO considering plant cells to be "microbiological products", it was held that plant cells did not come within the meaning of plant variety and are considered equivalent to bacteria and yeast, which are undoubtedly patentable as per the current practice of the EPO.[148] The concept

---

[148] However, the board was not convinced with the arguments that claimed plants were not the result of essentially biological process but the result of biotechnological process. The board also was not convinced that claimed plants did not

of "essentially biological process" was interpreted as "it depends on the totality of the human intervention and its impact on the results achieved to differentiate a microbiological process from that of essentially biological process."[149] The present invention for the preparation of hybrid plants is an essential modification of known biological process and it has got decisive impact on the desired result of hybrid population. Hence, it is not an essentially biological process and the exclusion mentioned in EPC Article 52(b) does not apply. Furthermore, it was held that the invention of a genetically modified cell is a result of microbiological process, which differs from an essentially biological process.

The board clarified what constitute a "microbiological process" and "products thereof" under EPC Article 53(b): processes in which microorganisms or their parts are used to make or modify products or in which new microorganisms are developed for specific uses are microbiological processes. The "products thereof", that is, products of microbiological processes encompass products made of or modified by microorganisms as well as new microorganisms as such. Hence, it was held that the present invention of plant cells and seeds being products of microbiological processes is potentially patentable under EPC. This case happens to deal in detail with the exclusion provisions under EPC. The board defined important concepts pertinent to patenting of biotechnological inventions. It was held that plant cells, being products of microbiological processes, were patentable. However, it was decided that the claimed plant, if it encompassed a plant variety, was excluded from patenting. In *Ciba Geigg*, exclusions were restrictively interpreted to allow patents on plants, but in *Plant Genetic Systems*[150] a more liberal construction of the exclusions was used to reject claims over plants. This led to the inference that plants produced as a result of genetic engineering were not patentable since the starting point of every such invention would be a plant variety.[151]

---

constitute plant variety within the meaning of Article 53(a) of EPC and rejected claims for plant on the ground that these claim encompassed plant varieties.

[149] See Sreenivasulu. 2011. *Intellectual Property Rights*.

[150] *Plant Genetic Systems v. Greenpeace* T356/93 Technical Board of Appeal 3.3.4 [1995] EPOR 357 (hereinafter *'Plant Genetic Systems'*).

[151] See generally Ramakrishna. 2003. *Biotechnology and Intellectual Property Rights*, p. 32.

In the meantime, the EU adopted a directive on biotechnology inventions in 1998.[152] The directive nullified the decision of the Plant Genetic Systems, in saying "inventions concerning plants shall be patentable" if the technical feasibility of the invention is not confined to a particular plant variety. Accordingly, now genetically modified plants are allowed patents even if they encompass a plant variety.[153] While in the *Novartis* case[154] the claims were on genetically modified plants to render them resistant to fungi and the Enlarged Board of Appeals ruled that as long as specific plant varieties were not individually claimed, claim for plants should not be excluded from patentability. Overturning Plant Genetic Systems' decision, the board stated that a claim that encompasses more than one variety was not excluded under Article 53(b) of EPC. It was viewed that in the absence of the identification of a specific plant variety in a product claim, the subject matter of the claimed invention was not directed to a plant variety or varieties within the meaning of Article 53(b) of the EPC. It was held that Article 53(b) exclusion was applicable only where the patent claimed a plant variety. Therefore, a patent cannot be denied only on the ground that the claim encompasses a plant variety. A claim that encompasses more than one variety is thus patentable. The decision however nullifies the decision of Plant Genetic Systems in holding a genetically modified plant patentable even when it encompasses plant varieties and brings the EPC in line with the EU directive on the biotechnological inventions. Now after the decision in *Novartis* clarifying that patents are granted through claims encompassing plant varieties and the adoption of the Directive stating that GM plants are patentable, plants produced through a biotechnological process are undoubtedly patentable in the EU.

Patenting of Transgenic Animals

Efforts were made in the EU to manipulate animals with the aid of biotechnology. In the process of evolution of biotechnology patent law,

[152] The directive on the legal protection of biotechnological inventions, 1998. The directive was fully implemented in all member states by July 2000.

[153] Article 4.2 of the directive.

[154] (1998) OJ EPO 149.

patents on transgenic animals were also granted along with patents on different biotechnological inventions such as microorganisms and transgenic plants. For the first time in the history of Europe, an animal in Harvard oncomouse was claimed for patent.[155] After it was decided by US Patent Office to allow patents on biotechnologically produced and genetically modified animals in 1988, it was the turn of the EU in 1990 to follow suit. The invention involved a transgenic mammal,[156] an animal susceptible to cancer, which was useful in testing anticancer drugs and technology. Inventors produced transgenic mice possessing foreign genes (oncogenes), which made them susceptible to cancer. The inventor claimed for a method for producing a transgenic nonhuman mammal susceptible to cancer and a transgenic nonhuman mammal containing oncogenes.[157] The method involved introducing an activated oncogene sequence into a nonhuman mammal at a stage no later than the eight-cell stage. Initially the patent was refused by the EPO examining division on the ground that it was against public order and morality and also that the invention encompassed an animal variety, which was excluded from the purview of patenting. EPC under Article 53(a) prohibits patent protection for invention the exploitation of which is contrary to public order and morality. At the same time, the convention under Article 53(b) prohibits inventions encompassing animal varieties from patenting.[158] The claim for nonhuman mammalian was considered very broad in the light of the invention describing only a method to produce transgenic mice.[159] In appeal, the EPO Technical Board of Appeal rejected the interpretation of EPC Articles 53(a) and (b) excluding the present animal invention from patentability. The board opined that examining division was wrong in refusing the patent application on the ground of exceptions to patentability under Articles 53(a) and (b). As

[155] T 19/90 (1990) O.J. EPO 476, Tech. Bd App; (1991) E.P.O, R.525, Ex. D.

[156] A mammal is a living being that breastfeeds its offspring. To put it otherwise, breast-feeding animals are called mammals. Human being is also a mammal, as human beings breastfeed their offspring.

[157] Sreenivasulu and Basavaraju. 2005. 'Biotechnology and Patent Law', pp. 1–2.

[158] Exceptions to patentability under EPC Article 53.

[159] See generally Jayashree. 2002. *Intellectual Property Rights*.

there was no specific mention of animal invention being patentable or not, it was inevitable that the board would interpret the existing provisions of the convention to decide the issue. The convention provided for exclusions for patenting and also exceptions to patentability. The board viewed that as per the language of the convention, animal inventions were not excluded from patenting. Therefore, the issue before the court was to decide whether the claimed invention did fall under any of the exclusions under the convention. In this background, the board happened to interpret and analyze exceptions to patentability under convention such as "order public," "morality," "animal variety," "essentially biological process," and "microbiological process."

While considering exception from patentability under EPC Article 53(a) on the ground of "order public", it was believed that all the complexities of biotechnological inventions needed to be considered carefully. First, the suffering of animals, and secondly, possible risks to the environment are to be weighed with the usefulness of the invention to the humankind. The board resumed the case to consider the claim again. The examining division viewed that "the development of new technologies is normally afflicted with new risks" and the humankind had to make this choice many times in the past. Therefore, the board opined that they should be careful weighing up of the risk with the positive aspects of such inventions. The board viewed that if the risk associated with the invention was higher than its benefits, the patent could be rejected on the grounds of "public order" and "morality". When higher life forms are involved in the new technology, it is not only the risk that must be considered, but also the possibility of harm to such higher life forms. In the case at hand, different interests were involved and required balancing: first, there was a basic interest of humankind for remedy of widespread and dangerous diseases, but on the other hand, the environment needed to be protected against the uncontrolled dissemination of unwanted genes and, moreover, cruelty to animals needed to be controlled. If the invention caused any kind of environmental imbalance or suffering to animals, denial of patents would be justified on the grounds of public order and morality.[160] The board opined that the usefulness of present

---

[160] See generally, Cornish. 1999. *W.R Cases and Materials on Intellectual Property*, pp. 62–7.

invention to humankind could not be denied. Cancer is one of the deadliest diseases of the world.[161] Any contribution to the development of new and improved anticancer treatment is therefore a benefit to the humankind and must be welcomed. The claimed invention contributed to a reduction in the overall extent of animal and human suffering. As there were no alternatives to animal testing in the given context, the invention could be allowed to serve the interests of humankind. It was viewed that the purpose of the invention was to provide animal test models, which would be used exclusively in the laboratory under controlled conditions by qualified staff. There would be no intended release of the invention into the environment. Such being the case, the risk of uncontrolled release would not arise except for the possible misuse or blatant ignorance on the part of the laboratory personnel carrying out the tests. This could not be a ground to deny patent grant. Concluding its discussion, the Examining Division stated that the claimed invention was not immoral or contrary to public order, as it was very much useful in testing anticancer drugs and technology. Usefulness to humankind outweighed animal suffering. While considering whether the invention constitutes "animal variety"[162] within the meaning of EPC(a), it was viewed that the claimed invention is a transgenic animal and did not constitute animal variety. Considering the fact that European patents were not granted for essentially biological processes,[163] the board found that the claimed process involved insertion of an oncogene into a vector in the present case plasmid, which was microinjected into the egg at an early embryonic stage, which was not possible through essentially biological processes.[164] The board viewed the claimed invention as a result of microbiological process as an exception to the essentially biological process and held the same as patentable.[165] The board, considering the arguments that the

---

[161] See Sreenivasulu. 2011. *Intellectual Property Rights.*

[162] See generally, Ramakrishna. 2003. *Biotechnology and Intellectual Property Rights*, p. 31.

[163] EPC Article 53(b) says essentially biological processes are not patentable.

[164] Bar of patenting of essentially biological processes does not cover microbiological processes. The present invention is a result of microbiological process wherein an oncogene is inserted into the mice at an early stage of eight cells.

[165] Microbiological processes are patentable under the convention in contrast to essentially biological processes.

claim for producing nonhuman mammal was too broad, as the invention claimed only a method to produce transgenic mice, held that a biological invention is considered sufficient to be granted a patent if it clearly indicates at least one way in which the skilled person could execute or produce it. The method need not disclose all the ways of practicing the invention on different mammals. Hence, it was held that the disclosure of a nonnatural method to produce transgenic mice was enough to meet the disclosure requirement, and eventually the patent was granted on oncomouse. For the first time in the history of European patent law, a patent was granted on a transgenic animal (mouse). This case interpreted and discussed in detail various provisions of EPC with relation to the patentability of biotechnology inventions. This decision cleared all doubts about the interpretation of EPC exclusion provisions in the light of patenting of biotechnology inventions. With this decision, now the EPC encompasses patents to transgenic animals produced through biotechnology.

Patenting of Human Genetic Material

For the first time in Europe, in the *Relaxin* case[166] the EPO happened to decide whether human genetic material was patentable or not.[167] The claim was for a gene coding for a hormone called relaxin. Relaxin is released in the body of a pregnant woman at the time of delivery to relax the body that labors and suffers pain. It was the first case where the patent for genetic material was claimed. The patent was opposed by contending that human genetic material did not constitute patentable subject matter under the EPC.[168] It was argued that patenting of human genes amounted to owning property rights in the human body, which was against ethical standards. However, the patent was granted by outweighing ethical objections regarding the claims to the patentability of human genetic material. It was held that patenting of human genes did

---

[166] (1995) *Official Journal of the European Patent Office* 388; (1995) E.P.O. R. 541.

[167] See generally, Beyleveld, Deryck and Roger Brownsword. 1998. 'Human Dignity, Human Rights and Human Genetics', *Modern Law Review*, Special Issue: Human Genetics and Law: Regulating a Revolution, 61(5): 661–80.

[168] EPC Article 52.

not go against ethics, as patenting of genes did not amount to patenting of human beings. Furthermore, it was opined that EPC did not exclude genes in isolated and purified form from patenting.

Following suit of the *Relaxin* case, human genetic material was again claimed for patent in *Biogen* v. *Medeva*.[169] The *Relaxin* decision did set the path for the patenting of human genetic material, as applicants in the present case claimed DNA molecules for patent. The invention was genetically engineered DNA molecules containing an insert needed to produce a crucial protein for a vaccine against the hepatitis B virus. The patent was based on experimental work done by Kenneth Murry of the University of Edinburgh in 1978. At that time, a fair amount was known about the hepatitis B virus, but the picture was far from clear. The patent was granted to Biogen.[170] At the instance of alleged infringement by Medeva, Biogen lodged a complaint where Medeva counterclaimed the invalidity of the patent on the ground of lack of inventive step as prior art suggesting the invention. Medeva alleged that Biogen used the known process and produced the known products. In support of their arguments, Medeva took recourse to a publication by an expert in the field on enabling the direct expression of a eukaryotic protein in a prokaryotic cell. The UK Patent Court, holding the patent as valid, viewed that the difference between prior art and the inventive concept is the decision and execution of the invention. Biogen's execution of the idea in producing genetically engineered DNA is its inventive step. In an appeal, the Court of Appeal overturned the decision of patent courts and held that Biogen used known and available techniques and materials in making the invention. The Court of Appeal opined that in the light of the prior art Biogen had not developed any new processes or new product and had not discovered anything new about the claimed process. The House of Lords upheld the decision of the Court of Appeal later. The biotechnology industry was not happy with the decision and stated that increasing the purity or quality of known products involved immense research effort and it should be acknowledged. Meanwhile, at the time when the House of Lords decided Biogen invention as not

---

[169] (1997) R.P.C. 1, HL.

[170] See generally, Cornish. 1999. *W.R Cases and Materials on Intellectual Property*, pp. 89–92.

valid, the EPO Technical Board of Appeal held the patent as valid, thereby providing some encouragement to the biotechnology industry. The Technical Board of Appeal accepted the contentions of the industry that increasing the level of purity or quality involved an immense research effort and the same deserved patent, holding the Biogen patent as valid. Furthermore, the board also considered the Biogen invention as a major achievement in the medical field as many people were working to discover drugs against hepatitis B. The decision of the EPO Board favored the biotechnology industry; but the decision of the House of Lords went against the desires of the industry. Both the decisions had come at the same time; however, the decision of the EPO Board had binding force throughout the EU and could be given more weightage. The decision of the board brought human genetic material such as genes and DNA within the purview of the patent law.

After the EPO Board declaring the Biogen patent on DNA molecule as valid, in *Novartis* the same EPO Board held that genetic material such as cells and parts were microorganisms and were thus eligible for patent protection.[171] Reiterating the *Biogen* decision, the board reiterated that human genetic materials were indisputably patentable throughout the EU. Hence, in Europe, living matter produced through biotechnology such as microorganisms, plants and related inventions, animal and related inventions, and human genetic material such as cells, genes, DNA, and parts thereof were indisputably patentable.[172] Although there were no specific legislations on patenting of biotechnology, the existing patent laws had been interpreted in such a way as to encompass patent protection to biotechnological inventions leading to emergence of biotechnology patent law within the existing patent law. The biotechnology patent law emerged with patenting of microorganisms. Later, plants were given patent protection, followed by patents on animals and at last granting patent protection to human genetic material. But there is a universal conviction among international community in prohibiting human being from the purview of patentability. Although law has also evolved with biotechnology, the international society is worried

---

[171] Sreenivasulu and Basavaraju. 2005. 'Biotechnology and Patent Law', pp. 1–2.

[172] See Sreenivasulu. 2008. *Biotechnology and Patent Law*.

about the regulation of the biotechnology industry. As biotechnology has been proved to be capable of manipulating any living being, it may move toward producing transgenic human beings, which is against the accepted ethical and moral standards of the society. Such inventions may give rise to public order problems. Although the international society is committed to prohibiting research on transgenic humans, it is worried about the developments in the industry that may violate ethical and moral standards of society. Meanwhile, seeing the enormous progress of the biotechnology industry and vehement demands for the proper legal protection of biotechnological inventions, the EU adopted a directive on the legal protection of biotechnological inventions.[173] The directive is a major breakthrough in the evolution of biotechnology patent law, as it provides for a blueprint on the legal protection of biotechnological inventions. The directive acknowledges and respects various decisions by EPO and its Technical Board in granting patents on nonnatural living beings produced through biotechnology. The directive gives statutory recognition to the various judicial decisions rendered on the protection of biotechnological inventions.

EU Directive on Biotechnological Inventions

The directive mandates member states to protect biotechnological inventions under their respective patent laws. It urges for modifications in the existing patent laws of the member states to give effect to the provisions of the directive. The directive defines concepts such as "biological material,"[174] "microbiological process,"[175] "essentially biological processes,"[176]

[173] European Union Directive on the Legal Protection of the Biotechnological Inventions 1998; the directive came into being in July 2000.

[174] The directive defines "biological material" to mean any material containing genetic information and capable of reproducing itself or being reproduced in a biological system.

[175] The concept is defined to mean any process involving or performed upon or resulting in microbiological material.

[176] It is defined as "a process for the production of plants or animals that is essentially biological and consists entirely of natural phenomena such as crossing or selection."

and "plant variety,"[177] which are very significant in the light of patenting of living beings such as microorganisms, plants, animals, and human genetic material produced through biotechnological processes. These definitions are important since the starting point of all the biotechnological inventions will be a biological material resulting from essentially biological processes.

Inclusive Zones of Biotechnology under the Directive

The directive does not specify the inventions that are patentable. There is no illustrative or exhaustive list of biotechnological inventions that are patentable under the directive. It states that inventions that are new, involve an inventive step, and are susceptible of industrial application shall be patentable even if they concern a product consisting of or containing biological material or a process by means of which biological material is produced, processed, or used. Inventions involving microorganisms, plants, and animals are patentable under the directive, as it declares that the biological material produced or isolated from its natural environment by means of a technical process is patentable.[178] As biological materials are not available in an isolated and purified form in nature, it is believed that isolation and purification involve an inventive step. At the same time, the directive states that a biological process that involves a human intervention possessing certain nonnatural characteristics is patentable. The directive implies that human intervention to a biological process makes it a nonbiological process, which is patentable under the directive.

Excluded Zones of Biotechnology under the Directive[179]

The directive states that certain biotechnology inventions are not patentable.[180] The directive considers ethical and moral standards and,

[177] The directive says that the concept of "plant variety" is defined by Article 5 of Regulation (EC) No. 2100/94, that is, Community Plant Variety Rights Regulation 2100/94. The aforementioned regulation defines a plant variety to mean "a plant grouping within a single botanical taxon of the lowest known rank."

[178] Article 3 of the European Patent Convention.

[179] Articles 4, 5, and 6 of the European Union Directive on the legal protection of biotechnology inventions.

[180] Article 4 of the directive.

at the same time, runs parallel to the EPC[181] and the TRIPS[182] in excluding certain inventions from patenting. The directive goes a step ahead from EPC and TRIPS and states that the human body and its various stages of formation and development are excluded from the purview of patent protection. Unlike the US Congress, the directive does not prohibit human cloning but it excludes human cloning from patenting.[183] The directive states that there shall not be any patent protection for the following:

1. Plant and animal varieties.
2. Essentially biological processes for the production of plants or animals. However, the aforementioned provision does not apply to "microbiological processes."
3. The human body at the various stages of its formation and development, and the simple discovery of one of its elements, including the sequence or partial sequence of a gene, cannot constitute patentable inventions.[184]
4. Inventions—the commercial exploitation of which would be contrary to public order or morality, such as:
    (a) processes for cloning of human beings,
    (b) processes for modifying the germ line genetic identity of human beings,
    (c) uses of human embryos for industrial or commercial purposes,
    (d) processes for modifying the genetic identity of animals, which are likely to cause them suffering without any substantial medical benefit to man or animals and also animals resulting from such processes.

---

[181] EPC Article 53(b).

[182] Article 27(3) of TRIPS.

[183] The US has enacted a specific legislation banning human cloning. See Human Cloning Prohibition of Act, 2003.

[184] However, an element isolated from the human body or otherwise produced by means of a technical process, including the sequence or partial sequence of a gene, may constitute a patentable invention, even if the structure of that element is identical to that of a natural element.

The directive requires its member states to harmonize their law related to the patenting of biotechnological inventions. Soon after the adoption of the directive, in *The Kingdom of Netherlands v. European Parliament and Council*,[185] the European Court of Justice, when asked to respond on the scope of the directive, held that the directive did not concern what happened before the grant of biotechnology patents and what type of research was done, but it was only concerned with the technical parameters that the inventions would have to fulfill to become eligible for patent. Furthermore, it was also viewed that the directive was not concerned about what happened after the grant of the patent, whether it would be used by the patent holder or it would be transferred to somebody. A little later in *Commission v. Italy*[186] the European Court of Justice was called upon to decide whether the directive and its implementation were mandatory and whether there existed any discretion of member states regarding the exclusive list of biotechnology inventions under the directive. The case was decided in the backdrop of Italy failing to implement the directive at the national level. It is necessary under the directive for all the member European states, including Italy, to protect biotechnological inventions under national patent laws through necessary adjustments in their existing setups. It was held that unlike Article 6(1) of the directive, which allows administrative authorities and courts of member states a wide discretion in applying the general moral exclusion on inventions whose commercial exploitation would be contrary to public order and morality, Article 6(2) allows member states no discretion in the implementation of the specific exclusion.[187] The directive encourages and protects various biotechnological inventions. The directive tries to balance public order, ethics, and morality with the development and innovations of biotechnology. Furthermore, the directive empowers the Commission's European Group on Ethics in Science and New Technologies to evaluate all ethical aspects of biotechnology.[188]

After going through the blueprint of the directive, it can be inferred that a biotechnology invention that does not fall under the excluding

---

[185] 2000 , ECR I-7079.
[186] ECR 1-5335, Case C-456/03.
[187] See Plomer and Torremans. 2009. *Embryonic Stem Cell Patents*, p. 187
[188] Article 7 of the directive.

provision of the directive is patentable on meeting requirements such as novelty, inventive step, industrial application, and written description. However, it may not be possible to describe a biotechnology invention in a written form. For this reason, deposit of the invention is prescribed to compensate the written description requirement. Given the typical nature of biotechnological invention, the directive states that where an invention involves the use of or concerns biological material that is not available to the public and that cannot be described in a patent application in such a manner as to enable the invention to be reproduced by a person skilled in the art, the description shall be considered inadequate for the purposes of patent law unless the biological material has been deposited with any international depository established under the Budapest Treaty.[189] The directive says that such deposit shall be made on or before the submission of patent application.[190] Public can have access to the deposited invention. If such deposited biological material ceases to be available, a fresh deposit shall be made.[191] Reportedly, since the early 1980s the EPO received around 15,000 patent applications associated with the field of biotechnology. Some 1,500 of these applications were related to transgenic plants, 600 to transgenic animals, and around 2,000 to DNA sequences. Based on the applicable law, the office has granted about 3,000 patents on biotechnological inventions until date.[192] The directive on biotechnology inventions came into being in July 2000, providing for comprehensive framework on the patenting of biotechnology inventions. The adoption of the directive puts an end to the fluctuating and contradictory decisions of the patent office and the EPO Technical Board in patenting biotechnology inventions. The directive streamlines the biotechnology patent law in providing for patenting of inventions ranging from microorganisms, plants, animals, and human

---

[189]   Budapest Treaty for facilitating for the deposit of inventions for the purpose of patent procedure, 1977.

[190]   Article 13 of the directive.

[191]   Article 14 of the convention.

[192]   Karmstra, Gerald, Mark Doring, Nick Scott Ram, Andrew Shead, and Herry Wixon. 2002. *Patents on Biotechnological Invention: The EC Directive*. London: Sweet & Maxwell.

genetic material.[193] It can be said that the evolution of biotechnology patent law in Europe is complete with the adoption of the directive. The journey of evolution of the biotechnology patent law in Europe started with the adoption of the EPC, which laid the foundation for patenting of living beings. In the meantime, the decisions of the patent office and the EPO Technical Board contributed toward the evolution of biotechnology patent law. In the end, streamlining the provisions of the EPC and various judicial pronouncements, the directive on biotechnology inventions provides for a comprehensive set of rules on biotechnology inventions. Now in the EU, the patent law has been harmonized with reference to the patenting of biotechnology inventions.[194]

On the whole, the evolution of biotechnology patent law can be traced back to the constitution of the US, which intends to foster the progress of science and technology. In the beginning, the judiciary of the US played an important role in the emergence of the biotechnology patent law through its innovative interpretation of the patent law. The liberal approach of the US judiciary was supported by the US Patent Office by granting patents on biotechnology inventions. The suit of the US judiciary and the patent office was followed by other nations. Initially, the EU was ahead of the US as patent claims on biotechnology inventions involving life were witnessed. However, it was the US judiciary that gave rise to the emergence of the biotechnology patent law through its judgment in *Chakraborty*. The decision of *Chakraborty* is a milestone and the starting point in the evolution of formal biotechnology patent law. The decision had far-reaching impact, as it influenced the

---

[193] The directive provides for patenting of different types of biotechnological inventions such as microorganisms, transgenic plants, transgenic animals, and human genetic material. On the other hand, it excludes few biotechnological inventions such as cloning of human beings and also the human body and its elements at various stages of its development.

[194] After the directive was adopted in order to implement in the EPC, the Administrative Council of the EU inserted a new chapter entitled "Biotechnological inventions" in part II of the EPC implementing regulation and amended the wording of Rule 28 (6) of EPC. See: Amended EPC implementing regulations, adopted on 16 June 1999, came into being on 1 September 1999.

entire world. Even today across the globe, the courts, while adjudicating matters involving biotechnology inventions, and the patent offices, while granting patents on biotechnology inventions, do follow the rationality of the *Chakraborty* case. The approach of the US is very liberal and their philosophy is that anything under the sun made by man is patentable. The evolution of biotechnology patent law is heavily influenced by this philosophy evolved by Thomas Jefferson, the author of American patent law. The EU was quick enough to follow this philosophy in liberally interpreting the existing patent laws to provide patent protection to biotechnology inventions. Later, the entire world, including India, followed suit by amending their local laws or bringing suitable legislations to give patent protection to biotechnology inventions. Meanwhile, the TRIPS agreement also postulates for patenting of inventions in all the fields of science, including livings beings produced through biotechnology. It is felt that the agreement has streamlined and uniformed the law related to biotechnology inventions. Although it faced strong oppositions and objections, the evolution of biotechnology patent law is complete and all biotechnological inventions such as microorganisms, plants, animals, and human genetic material including different methods to produce the same are patentable throughout the world.

Summary on IPR and Biotechnology

The canopy of intellectual property law is meant for promoting the cause of innovations and technology. Biotechnology innovations have been a cause and concern for revisiting and reviewing the cantors of intellectual property law. Different regions have reacted differently to the innovations of biotechnology, as they are quite unconventional and controversial. International societies, while formulating uniform norms, have tried their level best to address the concerns and reservations of various regions in this regard. In this chapter, an attempt has been made to analyze the development of intellectual property law in the US, India, and Europe. In India, intellectual property law does not allow patenting of certain biotechnology innovations including medical methods, agricultural methods, and innovations such as genetically modified plants, animals, and other living innovations except microorganisms.[195] Unlike the US and Europe, India is considered to be conservative in its approach toward conferring intellectual property protection to biotechnology innovations. Perhaps, the US has been flamboyant in this regard, while the EU has been cautious but progressive in terms of encouraging IPRs on biotechnology innovations. Biotechnology, as a tool, has great potential for overcoming the constraints to increasing production in various industrial sectors including agriculture and medicine, but for this, substantial investments are required. Scholars point out that having a strong intellectual property protection system will promote investment and research, which in turn would foster industrial production and the economy. It is argued that the current intellectual property system, through government acquisition, compulsory licensing, and other emergency and balancing means, is quite capable of ensuring that monopoly is under check. Therefore, the biotechnology industry is of the opinion that on the lines of the US and the EU, India can also encourage intellectual property and patent rights on biotechnologically produced and genetically engineered plants, animals, human genetic material, and other living innovations. However, the other side of the coin is that unlike developed nations such as the US and the countries of the EU, India is not in a position to expand its intellectual property and patent law canopy beyond a limit because of its economic, industrial, political, and social conditions. Besides, India, being an agrarian economy and a country with mass population, would have more food- and health-related needs than the developed nations, which may not be possible to fulfill if more intellectual property and patent rights are granted on biotechnology innovations that are capable of catering to the food and health needs of the mass population.

[195] See Section 3(h), Indian Patents Act, 1970.

5   Biotechnology, Trade, and Environment

D evelopment of trade in biotechnology has been a subject of intense policy debate all over the world, especially in the developing nations. Biotechnology offers new opportunities for significantly increasing the productivity of agriculture, reducing the cost of food production, and decreasing the environmental damage caused by agricultural practices.[1] It also provides opportunities for the commercial development of novel, suitable, and biologically and genetically compatible medicines. The United Nations Conference on Environment and Development (UNCED) in 1992 brought biotechnology to global attention, after having understood the wide range of business and trade avenues that it provides. While advocating the safe use of biotechnology, it was acknowledged that biotechnology offers new

---

[1]   Guriswamy, Laxshman D. and Jeffrey A. McNeely (eds). 1998. *Protection of Global Biodiversity: Converging Strategies*, p. 49. Durham, North Carolina, US: Duke University Press.

opportunities for global partnerships and can contribute to sustainable development.[2] There are diverse fields of business and industry where biotechnology provides great trade opportunities, though it needs to be regulated not only under the trade law regimes but under the legal regimes of environmental protection to keep a tab on the probable associated risks on the environment and public health. More than any other fields, it is the agriculture and medicine fields of biotechnology where there are huge trade and commercial opportunities. The wing of biotechnology responsible for innovation in the agricultural sector is known as agro-biotechnology, which is benefiting with great commercial yields. Significant claims have been made about the capacity of modern biotechnology to contribute to food security, in particular by increasing agricultural production in developing countries and enhancing the nutritional value of basic foods. The Green Revolution in India played an important role in achieving self-sufficiency in foodgrain production, along with infrastructural development. The relevance of agricultural biotechnology arises from the need for infusion of a new round of technological change in the agriculture sector. However, it is being widely accepted that the rapid adoption of biotechnology in agriculture without its necessary risk assessment could pose innumerable challenges to policy and governance.

International Trade Regime on Biotechnology Regulation

Perhaps the regulatory regimes governing international trade have been slow in adapting its rules to regulate the international trade and movement of biotechnology products.[3] Nevertheless, there is emerging global biotechnology trade regime to promote commercialization and global trade of biotechnology products. The trade regime is intending to regulate trade-related aspects of biotechnology while imposing certain obligations and restrictions to ensure safety and security of

---

[2] Ricardo Melendez, Ortiz and Vicente Sanchez. 2007. *Trading in Genes: Development Perspectives on Biotechnology, Trade and Sustainability*, pp. 199–200. London: Earth Scan.

[3] Evenson, Robert E. and Terri Raney (eds). 2007. *The Political Economy of Genetically Modified Foods*, p. 291. UK: Elgar Reference Collection.

the environment.[4] As development should not be at the cost of the environment, the emergence of global biotechnology trade regime is promising to balance trade and environment in biotechnology products and genetically modified organisms (GMOs). With the development and commercialization of GMOs around the world, an intense debate arose concerning the risks associated with their use. In the early 1990s, international regulation to govern the safe use of GMOs was in relatively advanced stages in the Organisation for Economic Co-operation and Development (OECD) countries, but practically nonexistent in developed countries. In countries such as the US, there is an established legal framework for the promotion of trade in biotechnology. The large-scale and commercial exploitation of biotechnology and GMOs was encouraged and promoted under the World Trade Organization (WTO) trade regime,[5] which controls trade relations of more than 170 countries at present. WTO is the principal body governing international trade, which came into being in 1995.[6] Much before the commercialization of biotechnology, the final rounds of WTO/General Agreement on Tariffs and Trade (GATT) negotiations began in 1986, which ended in 1994 and resulted in the formal establishment of WTO in 1995. Lack of international consensus in WTO made it impossible to include GMO trade in WTO trade regime. However, concerns regarding GMO trade and biotechnology were raised in November 2001 at the ministerial meeting of WTO in Doha, and it provided an opportunity to address international commercial policy conflicts pertaining to biotechnology. There were endeavors to constitute a regulatory framework to govern GMO trade, though WTO had been facing a crisis of legitimacy as its decision-making process was claimed to have been undemocratic and unable to address the concerns of the developing countries. The rules of international trade have been perceived to be slow in addressing the

[4] See Wuger, Danioel and Thomas Cottier (eds). 2008. *Genetic Engineering and the World Trade System: World Trade Forum*, p. 209. UK: Cambridge University Press.

[5] WTO has got membership of more than 150 countries where proposed biotechnology trade regime under the WTO system is applicable.

[6] See Somsen, Han. 2007. *The Regulatory Challenge of Biotechnology: Human Genetics, Food and Patents*, p. 107. UK: Edward Elgar.

challenges of commercialization of biotechnology, in particular agricultural biotechnology. However, international regulatory framework and WTO do not specifically provide for agricultural biotechnology or GMOs. This would augment the economic and political pressure on the developing countries impeding the development of an effective and viable legal system. The inequitable structure erodes the policy autonomy of the citizens of the developing countries. It is believed that the US and the European Union (EU) utilize bilateral pressures to dominate the developing countries through instruments such as technical assistance or food aid, as support of this kind does not come free of cost or without any strings. Technical assistance is provided on the condition that the recipient countries adopt regulatory policies that are favorable to the benefactors and food aid serves as an instrument to capture new markets for GM food products.[7] In particular, the donor countries have "captured" the governments of the recipient developing countries, with the result that aid initiatives largely serve the geopolitical and economic interests of the donors and the narrow interests of these governments. This capture, in fact, leads to a lack of effective democracy (that is, participation in and accountability of governmental decision-making processes) in recipient countries. Thus, efforts to ensure that international regulation responds to the needs and concerns of the citizens of these developing countries must therefore include the democratization of national governance frameworks, which is possible through adoption of meaningful institutional frameworks for public participation in biotechnology regulation.[8] The GMOs are capable of spreading to the limits of their ecological niche, oblivious to international boundaries, and once released into the environment the spread of a GMO would be difficult to arrest because on release they may cross international borders, spreading diseases by the airborne transport of spores. Hence, both the citizens and the environment of one country can be affected

[7] Kripke, Gawain. 2005. 'Oxfam, Food Aid or Hidden Dumping? Separating Wheat from Chaff', Oxfam Briefing Paper No. 71, available at https://www.oxfam.org/sites/www.oxfam.org/files/bp71_food_aid.pdf (last accessed 26 February 2016).

[8] Akech, J.M. Migai. 2006. 'Developing Countries at Crossroads: Aid, Public Participation, and the Regulation of Trade in Genetically Modified Foods', *Fordham International Law Journal*, 29: 265.

by a deliberate release originating in another country, thereby creating an international concern.[9] Since the ecological and geographic ranges of GMOs transcend political boundaries, the potential risks and the variation in current deliberate release regulations in individual countries illustrate the need for an international approach to regulate deliberate releases of GMOs. As a rule, GATT mandates non-discrimination, that is, equal treatment of identical products, among all contracting parties,[10] and it further requires national treatment, that is, equal internal treatment[11] of both imported and domestic products. Article 20 of the GATT permits countries to take necessary measures intended to protect human health and the environment, provided such measures are not applied in a manner that would constitute a means of arbitrary or unjustifiable discrimination between countries where the same conditions prevail, or a disguised restriction on international trade.[12] In practice, this means that environmental and health-related measures must be scientifically based and are no more trade-restrictive than necessary to meet their goals. If an otherwise legitimate import restriction is not based upon solid science, it would be ruled invalid by the WTO, thus limiting the ability of governments to enact protectionist trade measures under the guise of environment protection. In determining whether a given restriction is scientifically based, the WTO looks at existing scientific research and the relevant standards set by Codex Alimentarius Commission, the International Plant Protection Convention (IPPC), and other relevant international standards.[13] However, at the Uruguay Round in 1994, it was observed that Article 20 had some "gray areas" that needed to be resolved. For instance, Article 20 does not establish

---

[9] Kim, Judy J. 1993. 'Out of the Lab and into the Field: Harmonization of Deliberate Release Regulations for Genetically Modified Organisms', Fordham International Law Journal, 16: 1160.

[10] Most Favored Nation (MFN) Obligation under Article I: 1 of GATT.

[11] National Treatment Provision under Article: III of GATT.

[12] Howse, Robert and Petros C. Mavroidis 2000. 'Europe's Evolving Regulatory Strategy for GMOs—The Issue of Consistency with WTO Law: Of Kine and Brine', Fordham International Law Journal, 24: 317.

[13] Adler, Jonathan H. 2000. 'More Sorry than Safe: Assessing the Precautionary Principle and the Proposed International Bio-safety Protocol', Texas International Law Journal, 35: 173.

any criteria for determining whether measures to protect human health and the environment were necessary and no specific procedure was provided for settling disputes on such matters; the Agreement on Sanitary and Phyto-sanitary Measures (SPS agreement) and the Agreement on Technical Barriers to Trade (TBT agreement) were the result of this review process.[14] The WTO multilateral trade regime prescribes the SPS agreement and the TBT agreement to create specific trade exceptions[15] including those for trade in biotechnology and GMO. Let us discuss these agreements and international standards for regulating biotechnology trade in some detail.

Agreement on Sanitary and Phyto-sanitary Measures

The SPS agreement recognizes that it is sometimes unavoidable to restrict trade in order to protect human, animal, or plant life and health. While allowing member states to take the sanitary, that is, human and animal health-related, and phyto-sanitary, that is plant health–related, measures necessary to protect health, the agreement seeks to prevent unnecessary trade obstacles.[16] The SPS agreement requires the members to ensure that measures are "based on scientific principles and are not maintained without sufficient scientific evidence," and that these measures should be "based on an assessment appropriate to the circumstances taking into account risk assessment techniques developed by the relevant international organizations." The SPS measures are general procedural requirements that are to be followed by the contracting parties with the chief aim to prevent restriction on international trade disguised as health and safety measure. In particular, it requires the members to base their SPS measures on science and not to use them as disguised barriers to trade, and presumes that "measures which conform to international standards, guidelines or recommendations are necessary to protect human, animal or plant life or health." However, where

---

[14] Bentley, Philip. 2000. 'A Re-Assessment of Article XX, paragraphs (b) and (g), of GATT 1994 in the Light of Growing Consumer and Environmental Concern about Biotechnology', *Fordham International Law Journal*, 24: 107.

[15] Somsen. 2007. *Regulatory Challenge of Biotechnology*, p. 107.

[16] Ricardo Melendez and Sanchez. 2007. *Trading in Genes*, p. 218.

a member proposes to impose measures stricter than those established by international standards, it can do so only if it provides sufficient scientific justification for its proposed measures. In the latter scenario, the member is required to undertake a scientific "risk assessment" to evaluate the likelihood of adverse consequences.[17] The risk assessment must be based on an examination and evaluation of available scientific information, and will only justify the imposition of SPS measure if a "rational relationship" exists between the risk assessment and the measure.[18] Further, the SPS measure that passes this science test must not be more trade-restrictive than necessary, must be consistent with comparable regulations, and must be imposed without undue delay. The SPS agreement also requires members to maintain transparent SPS regulations and prohibits the use of control, inspection, and approval procedures as these are unjustified barriers to imports. The SPS agreement applies to premarketing approvals for novel food containing living organisms and requires labeling directly related to food safety, such as a label warning of potentially allergenic genes introduced in novel food, but not a "Contains GMOs" label for consumer information.[19] The SPS committee met for the first time during 31 October to 1 November 2001 to discuss the issues of GMOs, their security, and possible effects on human health and the environment.[20]

Agreement on Technical Barriers to Trade

The TBT agreement applies to deals with standards and technical regulations, such as packaging, marking, and labeling requirements, which are not promulgated for sanitary or phyto-sanitary purposes. This agreement would likely apply to technical barriers imposed on the import of

---

[17] See Victor, Marc. 2001. 'Precaution or Protectionism? The Precautionary Principle, Genetically Modified Organisms, and Allowing Unfounded Fears to Undermine Free Trade', *Transnat'l L.* 14: 295, 296.

[18] Akech. 2006. 'Developing Countries at Crossroads', p. 265.

[19] Eggers, Barbara and Ruth Mackenzie. 2003. 'The Cartagena Protocol on Bio-safety', *Journal of International Economic Law*, 3: 525.

[20] Chaturvedi, Sachin and S.R. Rao. 2004. *Biotechnology and Development: Challenges and Opportunities for Asia*, p. 304. New Delhi: Academic Foundation.

living modified organisms (LMOs), not to protect the environment or human health but to inform consumers or to protect a state's culture or economy.[21] Under this agreement, WTO members may not, unjustifiably or arbitrarily, discriminate against imports through their technical regulations and standards. In addition, technical regulations may not "be more trade-restrictive than necessary to fulfil a legitimate objective." Unlike the SPS agreement, the TBT agreement does not require scientific justification for any standards or technical regulations, as its scope "extends beyond measures that could justify risk on scientific assessment (such as professional licensing regimes)." Therefore the TBT agreement's risk assessment requirements seem to be broader and "much less rigorous" than those of the SPS agreement, making it easier for the regulators to justify a GM food-related measure.[22] The TBT agreement will often provide the relevant measure for judging labeling regimes for genetically engineered food. The range of legitimate objectives can include those mentioned in the GATT such as "public morals" and those that have been acknowledged in international law, which may include human rights law. For the regulation of trade in biotechnology and GMOs, it is perceived that the TBT agreement is very relevant and purposive, as it provides much-required international regulatory regime[23] on the labeling of GMOs and GM food. However, the members shall ensure that technical regulations are not prepared, adopted, or applied with a view to or with the effect of creating unnecessary obstacles to international trade.[24]

Codex Alimentarius Commission

The Codex Alimentarius Commission (henceforth Codex), an intergovernmental body, was created by the United Nations Food and Agricultural Organization (FAO) and the World Health Organization (WHO) in 1961. The purpose of the body was "to guide and promote

---

[21] Murphy, Sean D. 2001. 'Biotechnology and International Law', *Harvard International Law Journals*, 42: 48, 79.

[22] Akech. 2006. 'Developing Countries at Crossroads': 265.

[23] Chaturvedi and Rao. 2004. *Biotechnology and Development*, p. 294.

[24] Ricardo Melendez and Sanchez. 2007. *Trading in Genes*, p. 220.

the elaboration and establishment of definitions and requirements for foods, to assist in their harmonization and, in doing so, to facilitate international trade."[25] The basic objective of the Codex is to encourage the formulation and harmonization of food standards worldwide through consensus-building discussion and negotiation. It is responsible for managing a joint FAO/WHO Food Standards Programme. So far, the Codex, having the membership of 187 countries, has developed more than 200 food standards for commodities and more than 40 codes of hygiene and technological practice. The Codex garnered much recognition after SPS agreement entered into force. The Annexure A to the SPS agreement states: "for food safety, the relevant standards, guidelines and recommendations are those established by the Codex Alimentarius Commission[26] relating to food additives, veterinary drug and pesticide residues, contaminants, methods of analysis and sampling and codes and guidelines of hygienic practice." There is a relationship between the SPS agreement, the Codex, and the Cartagena Protocol on Biosafety. Since precautions are being taken to ensure the safety of human health and environment, it can be said that SPS reading with the Codex perhaps echoes the precautionary principle of the protocol.[27] The GM food safety issues are currently being considered by the Codex. There are different committees under the commission that are related to GM food and GMO. These committees are as follows:

1. Committee on General Principles
2. Committee on Food Labelling
3. Codex Committee on Food Import and Export Inspection and Certification Systems
4. Ad Hoc Intergovernmental Task Force on Food Derived from Biotechnology.

The Committee on Food Labelling is host to international discussions on the rules for food labeling, including biotechnology. In 1999, an Ad

---

[25] Stewart, Terence P. and David S. Johnson. 1998. 'The SPS Agreement of the World Trade Organization and International Organizations', *Syr. J. Int'l L. & Com.*, 26: 27, 41.

[26] See Ricardo Melendez and Sanchez. 2007. *Trading in Genes*, p. 72.

[27] See Somsen. 2007. *Regulatory Challenge of Biotechnology*, p. 110.

Hoc Intergovernmental Task Force on Food Derived from Biotechnology was established to develop standards for GM foods. The task force[28] discusses scientific evidence, risk analysis, and other legitimate factors relevant to the health of consumers and the promotion of fair trade practices in GM food. The status of the Codex has grown so much with the adoption of the SPS agreement that the Codex guidelines and standards are generally considered while adjudicating WTO trade disputes and concerns. The Codex, through recognizing the more precautionary approach, is also intimately connected to the WTO and its agreements.[29] The commission prescribed eight-step approval process for the GM food-related issues, deliberations on which are stuck at the third step only because of disagreements among the parties. The differences between the US and the EU[30] regarding the labeling of GM foods were quite evident in their submissions to the Committee on Food Labeling. The US is of the opinion that it would agree for mandatory labeling only when there is detectable presence of transformed DNA or protein that makes the biotechnology-derived food significantly different from the conventional food. According to the US, there ought to be significant difference between the two foods in terms of composition, nutritional value, and intended use. The US has been on the other side of labeling of foods derived from biotechnology, so it has been very particular in voicing its opposition to mandatory labeling of GM foods. However, the EU is of the opinion that there should be mandatory labeling for GM food or biotechnology-derived food irrespective of any detectable and significant differences.[31] The primary purpose of the EU food policy is to protect the interest of the consumers and allow

[28] Notified amendments to the Procedural Manual of the Codex Alimentarius Commission, 23rd Session, June–July, 1999, Codex Alimentarius Commission, Food and Agricultural Organization, United Nations, World Health Organization, Joint Office, Rome, Italy, available at ftp://ftp.fao.org/codex/Reports/Alinorm99/al99_10e.pdf (last accessed 26 February 2016).

[29] Evenson and Raney (eds). 2007. *The Political Economy of Genetically Modified Foods*, p. 37.

[30] Codex Committee on Food Labeling, 2001.

[31] Paarlberg, Robert L. 2002. 'The Contested Governance of GM Foods: Implications for U.S.-EU Trade and the Developing World', Paper No. 02–04, Weatherhead Center for International Affairs, Harvard University, Massachusetts, US, available at http://projects.iq.harvard.edu/files/wcfia/files/558_paarlbergwp02-04.pdf (last accessed 26 February 2016).

them to make informed choices about the food they consume including the GM food.[32] In July 2003, the commission adopted three risk analysis standards for food derived from biotechnology, including general principles for the risk analysis of such foods and more detailed draft guidelines for the food safety assessment of foods derived from recombinant DNA (rDNA) plants and microorganisms. These standards contain references to the tracing of products and food labeling as risk management tools, and an annex on the assessment of possible allergenicity. The Codex Committee on Food Labelling has been working on recommendations for labeling of foods derived from biotechnology.[33] It is expected that the committee would formulate those means of labeling of foods derived from biotechnology that would specifically inform the consumer about the biological, physical, and chemical properties of the foods for their informed purchase, use, and consumption.

International Plant Protection Convention

The International Plant Protection Convention (IPPC) was created in 1951 to "secure common and effective action to prevent the spread and introduction of pests and plants and plant products and to promote measures for their control."[34] Thus, the IPPC helps to coordinate regional and international quarantines and sets voluntary standards on pathogens and potential plant pests. The standards set in the convention are being used for regulation of agricultural biotechnology to ensure the safety of the environment and prevention of spread of pests and potential risks.

Convention of Biological Diversity

In 1992, the United Nations (UN) Convention on Biological Diversity (CBD) was signed by over 150 governments at the UNCED, popularly

---

[32] Official Journal of the European Communities, 2001b, C96E/254.

[33] Ricardo Melendez and Sanchez. 2007. *Trading in Genes*, p. 224.

[34] Food and Agricultural Organization of the United Nations, Conference, 29th Session, Revision of the International Plant Protection Convention, C 97/17, at 1 (18 November 1997), quoted in Stewart and Johansson. 1996. The SPS Agreement of the World Trade Organization and International Organizations: 41.

known as the Earth Summit, held in Rio de Janeiro, Brazil, where the world leaders agreed on a strategy to meet the world's needs without sacrificing a healthy, viable world for future generations under the principle of "sustainable development," which recognized the conservation of biological diversity as a "common concern of humankind" for the first time. Till date, CBD is the most popular international convention in the history with more than 190 countries having ratified and adopted it. Under CBD, biotechnology is defined as "any technological application that uses biological systems, living organisms, or derivatives thereof, to make or modify products or processes for specific use."[35] It explicitly acknowledges the precautionary approach,[36] which is highly recommended for biotechnology trade. The CBD is the first international legally binding instrument containing provisions on biotechnology.[37] During the preparations for the 1992 Earth Summit, the issue of biotechnology and biosafety surfaced in a number of working groups, especially among those responsible for negotiating and drafting Agenda 21 and the CBD. Agenda 21 embraces the possibilities of using biotechnology for sustainable development of biological and other resources, particularly in the developing states.[38] The arguments for and against biotechnology products as a tool for sustainable growth, and the need to explore the possibility of international regulation, were addressed in several chapters of Agenda 21 as well in the text of the CBD, specifically in Article 8(g) on in situ conservation and Article19 on handling of biotechnology and distribution of its benefits.[39] The CBD, meant for safeguarding the biological diversity, acknowledges that the release of GMOs (referred to as "LMOs" in the CBD) may have adverse effects on biological diversity. An exclusive protocol, Cartagena Protocol on Biosafety, on addressing the safety issues involved in biotechnology and use of biogenetic resources,

[35] United Nations Convention on Biological Diversity, June 5, 1992, art. 2, 31 I.L.M. 818, 823 (1992).

[36] See generally, Evenson and Raney (eds). 2007. *The Political Economy of Genetically Modified Foods*, pp. 293–9.

[37] Ricardo Melendez and Sanchez. 2007. *Trading in Genes*, p. 171.

[38] Francioni, Francesco and Tullio Scovazzi (eds). 2006. *Biotechnology and International Law*, p. 66. Portland, USA: Hard Publishing.

[39] Francioni and Scovazzi (eds). 2006. *Biotechnology and International Law*, p. 154.

was adopted in 2000. The IPPC created in 1951, the Codex that came into being in 1961, the SPS agreement through WTO in 1995, and the Cartagena Protocol that was adopted in 2000 while aligning together, all speak of ensuring the safety of human and plant health, food, biological resources, and the environment. These agreements together regulate the issue of trade in GMOs, and the nations have adopted these agreements at the local level by enacting suitable laws and legislations concerning the issue of trade in GMOs.

Understanding GMOs

One of the principal contributions of agro-biotechnology is the production of GMOs/GM crops or derived food products. GMOs are also known as LMOs that possess novel combination of genetic structure through biotechnology.[40] Genetic modification or engineering of crops involves taking genes holding a desired trait or characteristic from one species and inserting the same into another. To make the desired modifications, the scientists employ a technology called rDNA, in which they identify specific genes, make copies of those genes, and introduce the gene copies into recipient organisms.[41] The desired traits could be pest and disease resistance, longer shelf life, better growth, sustenance in adverse conditions, high yield, and so forth. Cartagena Protocol on Biosafety defines LMO to mean: "as any living organism that possesses a novel combination of genetic material obtained through the use of modern biotechnology."[42] Scientifically, this would include the use of in vitro nucleic acid techniques, including rDNA and "direct injection of nucleic acid into cells or organelles" or "fusion of cells beyond the taxonomic family" to overcome natural physiological reproductive or recombination barriers. The GMOs produced through the process of genetic engineering is one of the enumerable modes of application of agro-biotechnology

---

[40] Hocking, Barbara Ann. 2009. *The Nexus of Law and Biology: New Ethical Challenges*, p. 96, England: Ashgate Publishing Limited.

[41] Moyer, Thomas J. and Stephen P. Anway. 2007. Biotechnology and the Bar: A Response to the Growing Divide between Science and the Legal Environment, *Berkeley Technology Law Journal*, 22: 671.

[42] See Article 3(g) of Cartagena Protocol on Bio-safety 2000.

to ensure food security for the fast-growing population. Some of the inventions produced through genetic modification that could be called GMO include:[43]

1. a new tomato with increased solid content that improves the taste and texture of tomato paste and processed sauces;
2. a recombinant form of cow hormone that increases the production of milk;
3. soybeans that produce healthier oils;
4. rice and grains with greater nutritional content;
5. bananas and potatoes that provide doses of vaccines.

The multinational corporations involved in genetic engineering offer substantial, immediate, as well as long-term economic advantage over the conventional seeds, with no concomitant disadvantage, something for which no definitive, clear-cut, transparent, and firm evidence has ever been presented to the public.[44] One of the primary arguments advocating in favor of GMOs is the increase in food security. Food security is directly linked to agro-biodiversity, which is essential to promote resilience in farming and ecosystem services on farms, such as pollination, fertility, and nutrient enhancement, insect and disease management, and water retention, making for more productive farming.[45] Therefore, there is a direct link between agro-biotechnology and biodiversity. Although the former is being used to ensure the latter, at the same time, growth in agro-biotechnology is resulting in concerns regarding the impediment that it creates in maintaining biodiversity.

---

[43] Generally, GMOs are intended for a variety of goals, such as to improve the nutritional value of cereals by enhancing the presence of special nutrients or chemicals, increasing crop and animal productivity in ways that will lower costs, improve quality and abundance, provide for the protection of the environment and maintenance of natural habitats, increase resistance to pests and diseases, or the achievement of better growth in adverse conditions, such as drought or cold temperatures.

[44] Bhargava, Pushpa M. 2002.'GMOs: Need for Appropriate Risk Assessment System', *Economic and Political Weekly*, April 13: 1402.

[45] Cullet, Philip and Radhika Koluru. 2002.'Plant Varieties Protection and Farmer's Rights: Towards a Broader Understanding', *Delhi Law Review*, 24: 41.

The agricultural and environmental applications of biotechnology require the intentional release of GMOs into the environment to test the new organisms' effectiveness under actual conditions, rather than in laboratory simulations. These types of introduction of GMOs into the environment are commonly known as the deliberate release into the environment. However, there are various apprehensions regarding genetically engineered microorganisms, as once released into the natural environment, theoretically they may have a good chance to survive, multiply, and exchange genetic material, or hybridize with other microorganisms in the environment; such hybridization between GMOs and naturally occurring counterparts may create new hazards. For example, wheat that has been genetically modified to resist pesticides may pass its pesticide-resistant gene on to a weed, creating the risk of disrupting ecological cycles.[46] Some believe that genetic engineering is part of a "human siege on the natural environment" and "agriculture worldwide is at risk" from the introduction of genetically engineered crops.[47] The natural life-supporting system on the earth has an intimate connection with human beings and this system cannot be altered irreversibly in the name of development. It is essential to understand that the entire debate on GMOs and biodiversity is grounded in the recognition of the urgent need to ensure environmental safety. Science must be given impetus but not at the cost of health, biodiversity, and the environment. Agenda 21 of the UNCED states that there is a need for development of internally agreed principles on risk assessment and management of all aspects of biotechnology, which should build upon those developed at the national level. Only when adequate and transparent safety and border-control procedures are in place will the community at large be able to derive maximum benefit from and be in a much better position to accept the potential benefits and risks of biotechnology.[48] Whether GMOs have a positive or negative impact on the ecosystem, biodiversity, farmers' rights, and the health of the people of a nation

[46]   Kim. 1993. 'Out of the Lab and into the Field', p. 1160.

[47]   Graziano, Karen M. 1996. 'Comment, Biosafety Protocol: Recommendations to Ensure the Safety of the Environment', *Colo. J. Int'l Envtl. L. & Pol'y*, 7: 179, 185, 188.

[48]   Ricardo Melendez and Sanchez. 2007. *Trading in Genes*, p. 200.

is a debatable issue and forms the subject of discussion. Biosafety, or the prevention of potential risks from the products derived from genetic engineering, has emerged as a pivotal scientific and political concern worldwide for which an effective regulation on GMOs is perhaps an important key to preserving the rich biodiversity of countries such as India, where more than 70 percent of the population is dependent on agriculture for livelihood.[49] The US has made rapid progress in the commercialization of GMOs, as reportedly, two-thirds of the US farms host GM crops. The first GM product was released in the market in 1994,[50] and after two decades, Argentina, the US, and Canada were proactive in the commercialization of GM crops. By 2000, out of 43.1 hectare of GM crops planted worldwide 98 percent were from these countries. The US accounted for 68 percent, Argentina 23 percent, and Canada 7 percent. In 2003, in the US biotechnological crops soybean accounted for 80 percent, maize 38 percent, and corn 70 percent. As of today, in the US, 70 percent of the agricultural activities are based on GM crops. The EU had imposed moratorium on GM products until 1998, and even when these were approved to enter the market, it was with cautious and slow commercialization. Nations such as Argentina, Canada, and Brazil are in the favor of commercialization of GM crops, whereas India, South Africa, and other such developing nations are not in its favor, particularly when it comes to GM crops. In India, GM cotton has received approval for being made available in the market, and there is stiff resistance for further approval of GM crops. In 2011, when the central government intended to give marketing approval to Bt brinjal in India, there was strong internal resistance from various stakeholders, which forced the government to withhold the approval.

CBD on Trade in GMOs and Safety

The CBD obligates the convention members to consider the need for and modalities of a protocol setting out appropriate procedures, including, in particular, advance informed agreement (AIA) in the field of the safe

---

[49] Lianchawh. 2005. 'Biosafety in India: Rethinking GMO Regulation', *Economic and Political Weekly*, 24 September.

[50] Ricardo Melendez and Sanchez. 2007. *Trading in Genes*, p. 259.

transfer, handling, and use of any LMOs that may have adverse effects on the conservation and sustainable use of biological diversity.[51] For transboundary movement of LMOs, it is compulsory to provide information regarding the use and safety regulations of the GM products to the other party.[52] At this juncture, the developing nations had reservations regarding biotechnology products and their impact on the environment and expressed their concerns over the absence of proper regulatory regime and, therefore, were not in favor of trade in biotechnology. The CBD established three main goals: the conservation of biological diversity, the sustainable use of its components, and the fair and equitable sharing of the benefits from the use of genetic resources.[53] In pursuance of these goals, the Cartagena Protocol was adopted to govern the trade-related aspects of GMOs. However, at the time of its drafting, several negotiating blocs were formed, which gave divergent opinions on the approach that the protocol should take. The nations were divided into different groups calling for stricter, flexible, and cautious measures according to their respective needs. Ultimately, as a compromise between the different groups, the protocol was adopted by accommodating the interests and concerns of all the groups. The countries involved in the production of GMOs sought market access for their GM food products to maximize their comparative advantage. The countries that are not involved in the production of GMOs sought for stricter regulations for safety from the side effects of GM products. So, all nations of the world are not unanimous in their stand on GMOs. The EU has taken a diametric approach compared to the US. The crux of the US and the EU[54] trade dispute is a conflict between the goals of free trade in GM products and the protection of human health and the environment when there is considerable scientific uncertainty surrounding biotechnology. The EU adopted a process-oriented approach that is largely driven by the need

[51]   See Article 19 (3) of CBD.

[52]   Article 19(4) of CBD.

[53]   United Nations Conference on Environment and Development: Convention on Biological Diversity, Article 1, 5 June 1992, available at http://www.biodiv.org/ doc/legal/cbd-en.pdf (last accessed 26 February 2016).

[54]   For further discussion on the US/EU conflict, see Akech. 2006. 'Developing Countries at Crossroads', p. 265.

for precaution and advocated the prohibition of the release of GMOs into the environment and the commercialization of GM foods "until there is extensive evidence that they will not cause harm to humans, animals and the environment." The process-oriented regulations view the technique of genetic engineering itself as a risk and regulate the use of rDNA techniques, even if the end product is not a GMO. In the EU's view, the precautionary approach is particularly suitable because it enables regulatory authorities to take consumers' concerns into account. It maintains that the Cartagena Protocol should take precedence over international trade agreements, and has adopted a regulatory model that requires prior governmental approval before the release of the GMOs into the environment. The US, on the other hand, alleges that these measures amount to protectionism and thus violate international trade rules. While the US adopts a product-oriented approach to the regulation of GM foods, it is guided by the principle that there should be minimal oversight of food products that are generally regarded as safe.[55] They argue that conventional food products are considered as safe, and GM foods should therefore be judged by the same standards, since they do not differ in any substantial way from those developed through traditional plant breeding methods. According to the US approach, the objective, therefore, should not be to establish absolute safety measures, but to consider whether a GM food (ingredient) is as safe as its conventional counterpart. The underlying principle is that zero tolerance for potentially hazardous ingredients in food would result in few foods being marketed ever. It has thus been noted that the problem with the US laissez-faire regulatory system is that "even when a company does not meet its food safety responsibilities, the regulator only takes action after some harm has resulted."[56] There is a need to encourage harmonization of international regulations for the deliberate release of GMOs into the environment for the development of genetically engineered products, to promote international trade, and to protect human health and the environment with common safety standards, keeping geographic and policy

---

[55] Perhaps, the product-specific regulations are not concerned with the use of biotechnology techniques, but with the use of the GMO end products such as foods or pesticides.

[56] Akech. 2006. 'Developing Countries at Crossroads', p. 265.

differences among nations in view.[57] It is pertinent to note that with reference to trade in GMOs the north has always been the producer and the south has been the consumer. The issue of GMOs is surrounded by political debate and public opposition. Due to the lack of scientific certainty in matters relating to the safety and risk of introducing GMOs into the environment, some of the regulators have responded with a wide range of deliberate release regulations while others have left the area completely free of regulations and made self-regulation the motto. The political and economic pressures exerted by the developed nations over the developing countries such as India, which requires the latter to follow the policies of the former, thereby foreclosing the opportunity of public participation in the decision-making processes of the developing countries, is a matter of concern. The influence of the First World, which is the largest producer of GMOs, paralyzes decision-making in developing countries and thereby impedes their ability to address their unique concerns regarding trade in GM food products. While the GM revolution has the potential to help alleviate the problem of hunger in the developing countries, the provision of GM food aid from the First World nations in the current international regulatory environment may instead increase the developmental problems of developing countries, since food aid undermines local agricultural production systems.

Moving Toward Evolving Biosafety Regulation

The 1992 Earth Summit in Rio de Janeiro had seen arguments being lobbed for and against the potential of biotechnology products to promote sustainable growth and the need to regulate such products on an international scale. Such concerns had found outlet in Agenda 21 and Articles 8(g) and 19 of the CBD. The Article 8(g) states: Each contracting party shall, as far as possible and as appropriate, establish or maintain means to regulate, manage, or control the risks associated with the use and release of LMOs resulting from biotechnology which are likely to have adverse environmental impacts that could affect the conservation and sustainable use of biological diversity, also taking into account the

---

[57] See generally Evenson and Raney (eds). 2007. *The Political Economy of Genetically Modified Foods*, pp. 293–9.

risks to human health. The convention mandates the contracting parties to take legislative, administrative, or policy measures, to ensure effective participation in biotechnological research activities.[58] This mandate, in particular, is supposed to provide for effective participation of developing countries that provide the genetic resources in biotechnological research.[59] The contracting parties are expected to adopt practical measures to promote and advance priority access on a fair and equitable basis, especially in developing countries, to the results and benefits arising from biotechnologies based upon genetic resources provided by those contracting parties. Such access shall be on mutually agreed terms. Furthermore, there shall be set modalities of a protocol setting out appropriate procedures, including, in particular, AIA, in the field of safe transfer, handling, and use of any LMO that may have an adverse effect on the conservation and sustainable use of biological diversity. It is the responsibility of the contracting parties to provide required information about the use and handling of biological resources and LMOs. Conflict of interests between the nations need to be resolved for building consensus to evolve internationally and universally acceptable legal regulation for biotechnology.[60] One of the most significant outcomes of the CBD is lending an international dimension to the discussions on biosafety and developing a set of accepted parameters on future negotiations in spite of an absence of consensus on the fundamentals of issues relating to biotechnology. There is a "saving clause" that ensures that the protocol does not supersede existing international agreements. Though the US was not a party to CBD, it wielded considerable influence over the coverage of the protocol, an action that, according to many, was a desperate attempt on its part to protect its own thriving biotechnology industry. The EU was vocal about it being in favor of inclusion of the precautionary principle into the AIA procedures and was against the

---

[58] See Article 19 of the Convention of Biological Diversity, 1992.

[59] See generally Aim, Martin, Anatole F. Krattiger, and Koachim von Braun (eds). 2000. *Biotechnology in Developing Countries: Towards Optimizing the Benefits for the Poor*, pp. 39–67. New York: Springer.

[60] See generally Evenson, R.E. and V. Santaniello (eds). 2004. *The Regulation of Agricultural Biotechnology*, pp. 230–40. Wallingform UK: CABI Publishing.

said "savings clause," which it felt could jeopardize denial of imports based on the precautionary principle.

The Cartagena Protocol on Biosafety

The CBD acknowledged that the release of GMOs may have adverse effects on biological diversity. In response to these concerns, the governing body of the convention negotiated and adopted the Cartagena Protocol on Biosafety on 29 January 2000 to address the potential risks posed by cross-border trade and accidental release of GMOs, to ensure biosafety during the transfer, handling, and use of LMOs and GMOs.[61] The protocol was agreed upon in Montreal in January 2000 as the first multilateral environmental instrument to regulate the transboundary movement of LMOs.[62] It covers both organisms that are intended for release into the environment and those destined for human consumption. The evolution of international policy could be traced to the adoption of Cartagena Protocol on Biosafety.[63] The protocol has the

[61] During the first meeting of the CBD member states known as the Conference of the Parties (COPs) in 1994 at Nassau, a discussion was held on how an intergovernmental process could be set in motion to define the biosafety protocol. Based on the views of the potential risks of biotechnology and it was agreed that there was a need for an international instrument. It lead to setting up a group of experts on biosafety for suggesting a legal framework for regulating the use, transfer, and movement of GMOs in the light of potential risks associated with it. In 1995, at Jakarta, a second conference of the parties of CBD took place where negotiations to launch biosafety protocol took place. A biosafety working group was constituted to negotiate for the purpose of the biosafety protocol. The group conducted meetings during 1996–8 at various places including Denmark, Montreal, and Cartagena but could not arrive at consensus regarding the protocol. In February 1999 at Cartagena, Columbia, attempts were made to convince different groups including Miami Group, Like-Minded Group, and the European Group but the discussions were incomplete. Further, in November 1999 at Vienna another round of meeting took place where some progress was made. Further, at Seattle in December 1999, the positions of different groups in the negotiating camps were clarified.

[62] Ricardo Melendez and Sanchez. 2007. *Trading in Genes*, p. 165.

[63] See Evenson and Raney (eds). 2007. *The Political Economy of Genetically Modified Foods*, pp. 293–9.

history of a hard but heartening negotiating process of the first legally binding multilateral framework that seeks to minimize the risks from LMOs to biodiversity and human health.[64] At one end of the spectrum, concerns were being voiced by the developing nations about their limited institutional and financial capacity to evaluate risks and minimize any negative impact of LMOs on the ecosystem. Such concerns were further compounded by the fear that lack of national biotechnology regulations may encourage field trials of LMOs in the territories of the developing nations once such trials fail to obtain authorization from more stringent regimes prevalent in the developed countries. On the other hand, nations having relatively sophisticated regulatory regimes and owning large stakes in the international market for biotechnology products perceived LMOs as the ideal solution, especially in the agricultural sector, to various problems faced by the global community, in the form of an increase in productivity and advocating conservation and sustainable use of biodiversity. Such nations further argued that biotechnology products being advanced DNA recombination, the usual risks associated with LMOs can be controlled largely in respect of such products as opposed to those obtained from conventional genetic modification. They were not in favor of the risk assessment procedures supposed to be applied to a given LMO before it could enter a country. According to them, such procedures were nothing but an undue restriction on trade and a barrier to the advancement of biotechnological research. Whether CBD can function adequately as a framework to discuss the safety of GMOs on a multinational level was also pondered upon at much length. The negotiations saw the emergence of Articles 8(g) and 19 of the CBD, both of which signified a step forward, at least with regard to the measures that individual countries would adopt at their discretion in the area of biosafety;[65] at the same time, these provisions also sought to pave the way for future multilateral discussions on LMOs. Nor was the concept of free trade neglected. The approval procedures referred to therein were supposed to apply only to those LMOs that could have been perceived to have an adverse effect on biodiversity, thereby allowing the option of

---

[64] Ricardo Melendez and Sanchez. 2007. *Trading in Genes*, p. 153.

[65] See Wuger and Cottier (eds). 2008. *Genetic Engineering and the World Trade System*, pp. 208–9.

excluding other LMOs from the purview of the regulations and removing a barrier to their free trade in the process. Precisely, the protocol is
intended to create a uniform international procedure for regulating the
safe transfer of LMOs.[66]

General Principles and Scope of the Protocol

The very purpose of the Cartagena Protocol is to protect biological
diversity from the potential risks posed by LMOs.[67] It requires products
from new technologies to be based on the precautionary principle and
allows the developing nations to balance public health against economic
benefits. For example, a participating nation can prohibit imports of a
GMO in case of lack of adequate scientific evidence ensuring the safety
of the product. Exporters are further required to label shipments containing genetically altered commodities such as corn or cotton. The basic
objective of the protocol is set out under Article 1 of the protocol, which
intends to: contribute to ensuring an adequate level of protection in the
field of safe transfer, handling, and use of LMOs resulting from modern
biotechnology that may have adverse effects on the conservation and
sustainable use of biological diversity, taking also into account risks to
human health and specifically focusing on trans-boundary movements.[68]
Certain general provisions of the Cartagena Protocol applicable to all
parties are as follows:

1. taking legal, administrative, and other measures to implement
   the obligations;
2. ensuring that LMOs are managed in a manner that prevents or
   reduces risk;
3. protecting sovereign rights of countries over their territorial seas
   and exclusive economic zones;

[66] Somsen. 2007. *Regulatory Challenge of Biotechnology*, p. 103.
[67] "Modern biotechnology" has been defined in Article 3(i) of the protocol
to mean the application of in vitro nucleic acid techniques, or fusion of cells
beyond the taxonomic family, that overcome natural physiological reproductive
or recombination barriers and are not techniques used in traditional breeding
and selection.
[68] Article 1 of the Protocol, SCBD 2000.

4. preserving the ability to invoke more stringent actions than those set out in the protocol while being consistent with it, and in accordance with international law;

5. using appropriate international forums with competence in human health risk issues.

Precautionary Principle

The issue of biosafety has been given priority by the nations at both international and domestic levels. Various measures have been adopted to address the issue of biosafety regarding the use of hazardous substances and nonnatural things. Since GMOs are nonnatural living beings that would be designed by marshalling the physical and chemical properties of targeted living beings, the concerns of safety in such experimentation and the use of modified organisms for practical purposes raise certain important issues. Article 1 of the Cartagena Protocol[69] states that the objective of ensuring an adequate level of protection for safe transfer, handling, and use of GMOs will be undertaken "in accordance with the precautionary approach contained in Principle 15 of the Rio Declaration on Environment and Development." The precautionary principle[70] is the golden thread and the ethos of the protocol. The precautionary principle is one of the eight salient features of sustainable development and implies that "where there are threats of serious or irreversible damage", lack of full scientific certainty shall not be used as a reason for postponing cost-effective measures to prevent environmental degradation.[71] The precautionary principle says that on certain occasions, particularly where the costs of action are low and the risks of inaction are high, preventive action should be taken, even without full scientific certainty about the problem being addressed. In practice, this gives governments a fair amount of discretion in setting environmental policy. Despite the fact

---

[69] See Evenson and Raney (eds). 2007. *The Political Economy of Genetically Modified Foods*, pp. 293–9.

[70] Hocking. 2009. *The Nexus of Law and Biology*, p. 98.

[71] Com (2000) 1 on the Precautionary Principle (European Commission, 2000) as given in Bell, Stuart and McGillivray, Donald. 2006. *Environmental Law*, 6th edn, p. 60. Oxford: Oxford University Press.

that science may be incapable, at least in the foreseeable future, of providing definite answers about the potential benefits and hazards of GMOs, the appropriateness of the precautionary principle as a factor in the decisions regarding biotechnology remains a highly debatable issue.[72] The principle itself had arisen out of the acknowledgment of the risks of serious and irreversible environmental harm of waiting for incontrovertible scientific evidence of harm before taking any preventive action. The protocol permits the adoption of precautionary measures to restrict imports of such organisms when the risk they pose for the environment or health is deemed incompatible with the level of environmental protection and safety set by the importing state.[73] However, some people fear that precautionary approach would slow its development, increase the costs of products and techniques, and limit the introduction of beneficial crops and food.[74] Others argue that the sound-science argument itself was an excuse to limit the use of an established principle of international environmental law.[75] It is submitted that although quite a few

[72] See Wuger and Cottier (eds). 2008. *Genetic Engineering and the World Trade System*, p. 209.

[73] Francioni and Scovazzi (eds). 2006. *Biotechnology and International Law*, p. 18.

[74] Adlar, Jonathan H. 2000. 'More Sorry than Safe: Assessing the Precautionary Principle and the Proposed International Bio-safety Protocol', *Texas International Law Journal*, 35 (173): 190.

[75] Certain provisions of the Protocol further reflect the influence of the said precautionary approach:
   a) The preamble, reaffirming the precautionary approach contained in Principle 15 of the Rio Declaration;
   b) Article 1, indicating that the objective of the Protocol is in accordance with the precautionary approach contained in Principle 15;
   c) Articles 10(6) and 11(8), which collectively declare that the lack of scientific certainty due to insufficient relevant scientific information and knowledge regarding the extent of the potential adverse effects of an LMO on biodiversity, taking into account risks to human health, shall not prevent a party of import from taking a decision, as appropriate, with regard to the import of the LMO in question, in order to avoid or minimize such potential adverse effects; and
   d) Annex III on risk assessment, which notes that the lack of scientific knowledge or scientific consensus should not necessarily be interpreted as indicating a particular level of risk, an absence of risk, or an acceptable risk.

liability regimes could be traced to deal with GMOs such as a domestic instrument or an international binding framework, at the threshold it is required that nations, especially the developing nations, enact stringent laws and policy guidelines to protect their people, animals, environment, and biodiversity.

Advanced Informed Agreement

One of the fundamental provisions of the protocol is its AIA or Prior Informed Consent mechanism, which requires exporters to obtain the consent of the country of import before shipping LMOs to that country for the first time.[76] The AIA procedure is set out in Articles 7–10 and 12 of the protocol. It involves notifying the importing country[77] in advance about the GMO in order to ensure the importing country takes necessary precautions and other required risk assessments according to their local needs. A party seeking to export an LMO destined for "intentional introduction into the environment" must notify the potential recipient country of its intention through the AIA procedure as the national governments have a right to know what is coming across their borders and should have the opportunity to ensure that the importation of new plant or animal species will not cause ecological harm. The potential importing country must then decide whether to permit the importation of the LMO. It is mandated under the protocol that the importing country should base its decision upon risk assessments carried out in a "scientifically sound manner."[78] The protocol requires the importing party to base its import decision on a scientific risk assessment "to identify and evaluate the possible adverse effects of living modified organisms on the conservation and sustainable use of biological diversity, taking also into account risks to human health."[79] The party of import then has 90 days to acknowledge receipt of the notification and inform whether it

[76] Articles 7–10 Cartagena Protocol on Biosafety available at https://bch.cbd.int/protocol (last accessed 26 February 2016).

[77] Ricardo Melendez and Sanchez. 2007. *Trading in Genes*, p. 174.

[78] Article 15 (1) Cartagena Protocol on Biosafety available at https://bch.cbd.int/protocol (last accessed 26 February 2016).

[79] See Safrin, Sabrina. 2002. 'Treaties in Collision? The Bio-safety Protocol and the World Trade Organization Agreements', *American Journal of International Law*, 96: 606.

intends to proceed with the procedure established by the protocol or according to its domestic regulatory framework. The protocol envisages a risk assessment to be carried out for all the decisions. Within 90 days of notification, the party of import must inform the notifier that it will either have to wait for written consent or may proceed with the import without written consent. If the verdict is to wait for written consent, the party of import has 270 days from the date of notification to decide to: either approve the import, adding conditions as appropriate, including conditions for future imports of the same LMO, or prohibit the import, or request additional information, or extend the deadline for response by a defined period.[80]

The party of import must indicate the reasons on which its decisions are based (unless consent is unconditional). A party of import may, at any time, in light of new scientific information, review and change a decision. A party of export or a notifier may also request the party of import to review its decisions.

Limitations of Advanced Informed Agreement

Within the provisions on AIA,[81] the protocol allows for a fair degree of flexibility. For instance, parties may proceed according to their own domestic regulatory framework, adopt simplified procedures, or enter into bilateral and regional agreements as long as these are consistent with the objective of the protocol. Alternatively, the potential importing country may require the exporter to conduct the risk assessment. The AIA procedure does not apply to LMOs in transit or destined for contained use, for example, in industrial production of enzymes or in fermentation processes. Further, it does not apply to LMOs intended for direct use as food, feed, or processing (the so-called LMO-FFPs), for which a different procedure has been provided wherein a country intending to

---

[80] Cosbey, Aaron and S. Burgiel. 2000. 'The Cartagena Protocol on Bio-safety: An Analysis of Results', International Institute for Sustainable Development, Canada, available at http://www.iisd.org/pdf/biosafety.pdf (last accessed 26 February 2016).

[81] See generally Somsen. 2007. *Regulatory Challenge of Biotechnology*, p. 104.

export LMO-FFPs is merely required to inform the potential recipient country of its decision through a "Bio-safety Clearing-House."[82] It is to be noted that most of the GM export consists of FFPs and by taking the same out of AIA framework, the entire effort of regulating trade related to genetically modified products becomes useless.[83] Similarly, the AIA procedure does not apply to LMOs transiting third states or destined for contained use (for example, vials for scientific research). According to Article 4, the protocol can be applied to only those LMOs that may have an adverse effect on the conservation and sustainable use of biological diversity or on human health. According to Article 7(1) of the protocol, it could also be applied to LMOs intended for intentional introduction into the environment. However, LMOs acting as pharmaceuticals that have already been addressed by other international organizations do not fall within the purview of the protocol's provisions on trans-boundary movements.[84] The crux of the protocol is the conduct or analysis of risk assessments as provided under Article 15 to inform regulatory decision-making, under the AIA procedure, on whether to grant or refuse the entry of the aforementioned LMOs. Under the protocol, Articles 8, 9, 10, and 12 lay down the entire AIA procedure consisting of notification, decision, and review processes.[85] Under CPB, the onus lies on the

[82] Akech. 2006. 'Developing Countries at Crossroads', p. 265.

[83] See Wuger and Cottier (eds). 2008. *Genetic Engineering and the World Trade System*, p. 209.

[84] While the AIA provisions do not come under its scope, LMOs in transit under Article 6 (1), LMOs destined for contained use in the party of import as per the Article 6(2), LMO-FFPs according to the Article 7(2), and LMOs identified by the meeting of the parties to the Protocol as being unlikely to have any adverse impact under Article 7(4).

[85] While Article 19 requires the identification of a competent national authority to administer the provisions, with Article 7 demanding compliance with the same by the party of export before the LMOs can be introduced into the environment of the party of import. Article 21 seeks to deal with the treatment of confidential information supplied as part of applications for trans-boundary movement of an LMO. Article 22 speaks of certain capacity-building measures to assist such countries to improve their so-called "bio-safety capacity." Last but not least, the socioeconomic considerations, as envisaged in Article 26, make a significant contribution toward the scope of AIA.

country reviewing the application for LMO release to prove that based on the information available, the said LMO may pose unacceptable risks to the environment or to human health. Most of the developing nations lack technical expertise for such evaluation.

Biosafety Clearing-House

Article 20 of the Cartagena Protocol provides for the establishment of a central Biosafety Clearing-House under Article 18(3) of the CBD in order to "facilitate the exchange of scientific, technical, environmental and legal information on, and experience with, living modified organisms" wherein the importing country is to provide the Biosafety Clearing-House with information about the crop, thereby demonstrating its acceptance of the crop and facilitating the accumulation of technical data about GMOs.[86] The protocol establishes an Internet-based Biosafety Clearing-House in order to facilitate the exchange of scientific, technical, environmental, and legal information on, and experience with, LMOs and to assist in the implementation of the protocol.[87] This multilateral information exchange mechanism, Biosafety Clearing-House, primarily pertains to the living modified products concerned with food, feed, and process.[88] It was implemented in a phased manner; the first meeting of the parties approved the transition from the pilot phase to the operational phase, and adopted modalities for its operations.[89] As far as the subject matter of liability and redress under the protocol is concerned, the issues seem to be quite straightforward, if hard to settle. The question is whether, and in what form, to create a liability and redress mechanism for any damage resulting from the trans-boundary movements of LMOs. In some form, this would involve the exporter or an insuring agent to pay for damages resulting from the import of its product. In this context, it should be mentioned that the threat of

---

[86] Akech. 2006. 'Developing Countries at Crossroads', p. 265.

[87] According to Aricle 20 of CPB, the Bio-safety Clearing-House will operate as part of the Clearinghouse mechanism that has already been established under Article 18 of CBD.

[88] See Ricardo Melendez and Sanchez. 2007. *Trading in Genes*, pp. 174–5.

[89] Decision BS-I/3, SCBD 2004.

costly mandatory segregation made the provision on documentation and identification the last to be settled in the Montreal negotiations. Those in favor of labeling and segregation found the result disappointing owing to the lack of any possibility of hard rules developed within two years after the protocol's entry into force.[90] While the Miami Group[91] had fought hard for this timeframe, arguing nonfeasibility of segregation, others argued that the growth trend for LMO in world trade would mean that within two years, market realities would anyway render the issue of segregation a moot point. A party that takes a final decision regarding the domestic use of an LMO that may be exported for food or for processing is obliged to inform the other parties via the Biosafety Clearing-House within 15 days.[92] Information that needs to be provided includes details about the producer and the LMO, as well as a risk assessment report, and the other parties may request additional information and take their own decision on the import of the LMO-FFP through their domestic regulatory framework. While the protocol leaves the regulation of the food and feed sector to the discretion of the parties, it also sets forth a procedure that may be used by developing countries and countries with economies in transition that have not yet enacted a national regulatory framework; as such countries may declare through the Biosafety Clearing-House that they will take decisions on imports of LMO-FFPs within less than 270 days, based on a risk assessment, and the CPB suggests that such a declaration would oblige exporters to refrain from exporting, pending a final decision.[93]

[90] See generally Evenson and Raney (eds). 2007. *The Political Economy of Genetically Modified Foods*, pp. 293–9.

[91] Miami group of nations comprising Argentina, Canada, and Chile had their own reservations about the Cartagena protocol which they spelt out during negotiations. See Cartagena Protocol. September 2003. 'A record of the negotiations', Secretariat of the Convention on Biological Diversity, World Trade Centre Montreal, Canada, available at https://www.cbd.int/doc/publications/bs-brochure-03-en.pdf (last accessed 27 June 2016).

[92] Article 11 of Cartagena Protocol on Biosafety, available at https://bch.cbd.int/protocol (last accessed 26 February 2016).

[93] Eggers and Mackenzie. 2000. 'The Cartagena Protocol on Bio-safety', p. 525.

EU Directives on GMOs and GM Foods

It is observed that due to political, administrative, and cultural differences that have defined the use of biotechnology, the nations have formulated regulations in their respective domestic frameworks of regulating trade in biotechnology. The EU adopted an exclusive directive on the deliberate release of GMOs into the environment.[94] The directive intends to regulate the release of GMOs into the environment for ensuring safety and security of the environment.[95] It is interesting to note here that the directive was adopted after the adoption of Cartagena Protocol on Biosafety by the members of the CBD. The EU always had different view on the matters of GMO trade and their impact on the environment. There have been attempts to adopt a stricter and more vibrant regime on the regulation and deliberate release of GMOs into the environment. Although the directive is only applicable to the member states of the EU, it does provide for the model framework on the regulation of trade issues in GMOs in addition to the Cartagena Protocol on Biosafety. The public perception in Europe is also not in favor of GMO trade. Extensive public surveys have been carried out in the EU member states on biotechnology-related issues. Public awareness on biotechnology is quite high in Europe, and seemingly a majority of the consumers are against biotechnology products and do believe that they could harm the environment and public health. People in Europe seem not to be in favor of giving control over the quality and flavor of the foods in the hands of GMO-producing companies, as food is seen as from the humanistic, aesthetic, national, and cultural identity.[96] In 2003, the EU[97] adopted a regulation for regulating the market placement of GMOs and GM foods.[98] It is a regulation for the placement in the marketplace of GMO food or products containing or consisting of GMOs.[99] In the same year, another regulation was adopted for ensuring traceability and labeling of GM products and GM foods.[100] It is worth mentioning here that there

---

[94] Naidu, David. 2009. *Biotechnology and Nanotechnology*, p. 168.

[95] Directive 2001/18 of 12 March 2001 (OJ L/106/1, 2001).

[96] Ricardo Melendez and Sanchez. 2007. *Trading in Genes*, p. 59.

[97] See Hocking. 2009. *The Nexus of Law and Biology*, p. 105.

[98] EC regulation 1829\2003.

[99] Naidu. 2009. *Biotechnology and Nanotechnology*, pp. 172–5.

[100] EC Regulation of 1830\2003.

was complete ban on GMOs and GM foods in the EU until 2001. Once the ban was lifted, the EU adopted the special directive for regulating the deliberate release of GMOs into the environment and regulations on market placement of GMOs and for ensuring traceability and labeling of GMOs.

Judicial Responses Regarding Disputes on Biotechnology Trade and Development

The international legal regime on biotechnology is, in general, very patchy and fragmented. Particularly, on the issue of trade in biotechnology and GMOs that are considered to have an impact on the safety of the environment and human health, the regulatory framework is extremely inadequate. The nations have been skeptical not only about the potential of biotechnology as such but also regarding the risks associated with the potential use and trade in biotechnology on the environment and health. Similarly, the nations are quite confused about the adoption of internationally acceptable principles and practices regarding the regulation of biotechnological trade. Law courts including International Court of Justice, Law of the Sea Tribunal, European Court of Justice (ECJ), and WTO Dispute Settlement Body all are rendering quite contrasting and contradictory decisions and guidelines. No universally acceptable judicial decisions or guidelines have been adopted in this connection. A study of the various judicial pronouncements leaves the impression that the law in this regard is yet to be finalized and provide for an established principle or approach on the most controversial issue of GMO trade. At the international level, there is lack of development of legal jurisprudence or case law development regarding the trade in GMOs. However, the EU provides for exception to this situation as the ECJ decided upon a number of disputes related to the issue of trade in GMOs.[101] Alongside, the WTO dispute settlement body has also delivered certain decisions in this regard. It can be said that these trade dispute and the inferences of the judicial bodies have contributed immensely to the formulation of a viable regulatory structure for GMO trade. These judicial decisions also highlight how deep and wide are the concerns involved in the trade

---

[101] Francioni and Scovazzi (eds). 2006. *Biotechnology and International Law*, p. 30.

in GMOs. They also display how countries are deeply divided over the issue, in particular the US and the EU.[102] It is worth conducting a study and analyzing how trade disputes and corresponding judicial decision have shaped and structured the current approach of the nations toward GMO trade. Let us discuss few important disputes and their judicial resolutions toward contributing the required legal jurisprudence.

Beef Hormone Case[103]

This was perhaps the first notified and widely speculated trade dispute involving GMOs. The involved parties in the dispute, the US and the EU, add further to its popularity. The different approaches adopted by the US and the EU regarding biotechnology trade were quite visible in this case. The EU imposed a ban on the illegal use of diethylstilbestrol (DES) in veal production. DES was found in baby food made from veal, and cases of children born with birth defects due to exposure to DES were reported from other places in Europe. This caused an alarm among the European consumers over the possible negative health effects of using hormones in livestock production. The "DES scare" made the entire Europe suspicious of the use of hormones in livestock production, and fearful of the potential harmful health effects of these practices. Such apprehension over the use of hormones in meat production led the EC to adopt the Nielsen Report, which approved the Commission's proposal of ban on the use of all hormones in livestock production unless the hormones were administered for therapeutic purposes. Based on the Nielsen Report, the EC adopted the ban on hormones in livestock production.[104] The directive dealt with five of the six hormones at issue, and the EC directed the commission to provide a report on the scientific assessment of the harmful effects of the five hormones. This led to the formation of the Lamming Report, which concluded that most of the

---

[102] Ricardo Melendez and Sanchez. 2007. *Trading in Genes*, p. 72.

[103] 'EC Measures Concerning Meat and Meat Products (Hormones) Complaint by the United States', WT/DS26/R/USA, 18 August 1997. Hereinafter referred as the Panel Report.

[104] Council Directive 81/602/EEC of 31 July 1981 concerning the prohibition of certain substances having a hormonal action and of any substances having a thyrostatic action.

hormones did not present any health risks when used under appropriate conditions as growth promoters in animals, but control programs for the use of these hormones was imperative, and additional scientific investigation was needed to assess the health effects of some of the hormones. Later, the EC introduced two directives. The first directive banned the use of all six hormones[105] for growth promotion purposes,[106] and the second directive imposed the ban on import of meat and products produced with the hormones in question.[107] However, certain objections were raised by both the EU members and certain non-European producers. Three member states, namely Belgium, Ireland, and the UK, questioned the need for a total ban on the use of hormones in livestock production. These objections were later echoed by non-European countries including Argentina, Australia, Canada, New Zealand, South Africa, and the US. Ultimately, the US filed the petition before the WTO Dispute Settlement Body on the issue of EC hormone ban alleging that there was no legitimate basis for the ban under various provisions of SPS agreement, TBT agreement, and GATT.

Claims by Both Parties

According to the US, the EU ban on growth hormones was not based on any scientific justification, which is a requirement under SPS rules. It was also claimed that the EU had failed to perform required risk assessments of the dangers posed by the hormones before implementing the ban; and the ban was meant to protect the EU cattle industry and not really based on health dangers. In the end, it was argued by the US that the ban constituted a disguised restriction on international trade. On the one hand, the US believed that consumer concerns should not play

[105] There are essentially six hormones that form the basis for the Commission's action and the subsequent GATT and WTO conflict. Three occur naturally in both cattle and humans (estradiol-17b, Tstosterone, and progesterone), and three are synthetic hormones (trenbolone, zeranol, and melengestrol). Acetate (MGA)) that mimics the action of the natural hormones.

[106] Council Directive 85/649/EEC of 31 December 1985 prohibiting the use in livestock farming of certain substances having a hormonal action.

[107] Council Directive 88/146/EEC of 7 March 1988 prohibiting the use in livestock farming of certain substances having a hormonal action.

a significant role in allowing a country to impose a ban that restricts international trade if it is not scientifically justified. On the other hand, the EU argued that further scientific study was needed due to inadequate scientific data on the safety of growth hormones, thus justifying EU's historical use of the "precautionary principle," and its different approach to risk assessment than the US. The EU stated that before executing the use of such hormone products it wanted to conduct further risk assessment studies, and that the EU believed in maintaining a higher level of protection for its consumers and the present ban was in keeping with these higher standards and it did not violate the SPS agreement.

Panel Report and Appeal

After studying the facts of the case and the arguments of the parties to the dispute, the panel concluded by stating that the EU's prohibition on beef hormones violated the SPS agreement because the ban was not based on a scientific risk assessment.[108] The panel attempted to define "risk assessment" under the SPS agreement.[109] The SPS agreement empowers the member states to adopt certain measures to protect health and to ensure environment safety on prescribed scientific assessment even though the same amount to trade restrictive measures. The panel agreed with the US claim that the EU standards were "arbitrary or unjustifiable distinctions which were discriminatory in nature and were basically a disguised restriction on international trade." The Dispute Settlement Body thus recommended that EU should adopt its policies in compliance with the SPS agreement. Otherwise, such actions of prohibition of beef hormones would amount to not only violation of the SPS agreement but also the broader trade obligations under the established WTO framework. The Appellate Body upheld the panel's finding that the beef hormone ban by the EU violated the SPS agreement.[110] The Appellate

---

[108] See generally Somsen. 2007. *Regulatory Challenge of Biotechnology*, p. 110.

[109] See Ricardo Melendez and Sanchez. 2007. *Trading in Genes*, p. 70.

[110] Appellate Body Report, European Communities: Measures Concerning Meat and Meat Products (Hormones), WT/DS26/AB/R (16 January 1998).

Body noted that "risk assessment is not only risk ascertainable in a scientific laboratory but is also risk in human societies as they actually exist." Although the appellate panel rendered the EU's ban on beef hormones as not acceptable, its findings showed that the panel might consider cultural variables as well as scientific data while deciding whether a member nation had carried out appropriate scientific risk assessment. Therefore, the question of cultural and societal differences influencing trade matters remains a matter of considerable concern. The case showed how judicial bodies implement SPS agreement and address the issues pertinent to trade obligations of member states under the WTO. The panel did not accept the traditional and precautionary approach of the EU. However, both parties tried their best efforts to come to a negotiation, with EC buying non-hormonal meat from the US and, on the other hand, the US decision to relieve or remove the sanctions earlier imposed by it. Further, it is viewed that following the beef hormone case, a court may conclude that labeling requirement as prescribed under TBT agreement for genetically engineered foods cannot be based on the objective of protecting human health since those foods do not significantly differ from conventional foods in their nutritional value.[111]

Gabcikovo-Nagymaros[112]

This was one of the initial cases where Hungary and Slovakia were contesting against each other regarding the prior information and risk assessment test in the case of GMOs. Like in the case of Beef Hormone, in this case too, the precautionary approach was not insisted upon by the International Court of Justice. The rationale adopted in the present case is similar to that of the WTO dispute settlement body in Beef Hormone. Advanced information procedure or the prior informed consent was not considered necessary within the framework of the GMO trade regulation. The court refrained from pronouncing the legally binding nature of the precautionary approach as either a custom or a general principle.[113]

[111] See Evenson and Raney (eds). *The Political Economy of Genetically Modified Foods*, p. 306.

[112] ICJ Rep 7, 1997, part at 42, Para. 54.

[113] Francioni and Scovazzi (eds). 2006. *Biotechnology and International Law*, p. 19.

However, it was considered that parties shall be duty-bound to cooperate and provide continuous impact assessment of technological risks associated with the biotechnology trade, which could impact the health and safety of the environment. The rationale of the Beef Hormone has been reiterated and upheld by the International Court of Justice in this case.

Australia Salmon Case[114]

The WTO dispute settlement body, while talking about the rights of the parties to the GMO trade dispute, held that establishing appropriate level of protection in case of GM products is a prerogative of the WTO member state and nothing can preclude the same. Therefore, based on sufficient evidence of potential risk, member states can adopt required measures to ensure their local interests, including protection of health and the environment. However, sufficiency of the evidence regarding the potential risk under the SPS agreement is a matter of concern for the state taking required legal measures in case of GM trade.

Japan Varieties Case[115]

The WTO dispute settlement body attempted to address the issue of the precautionary principle within the SPS agreement. It was viewed that nations may adopt precautionary approach in the case of unavailability of proper information and scientific assessment in trading of the GM product. The precautionary approach could be adopted for a more objective assessment of the risk. It seems even before the adoption of the biosafety protocol, the basic principle under the protocol, namely the precautionary approach in case of GM product, was suggested by the WTO dispute settlement body by reading the precautionary principle within the SPS agreement.

[114] WTO Dispute Settlement Appellate Body (AB) decision: WT\DS18\ AB\R, 20 Oct, 1998, available at https://www.wto.org/english/tratop_e/dispu_ e/cases_e/ds18_e.htm (last accessed 29 February 2016).

[115] WTO Dispute Settlement Appellate Body (AB) decision: WT\DS76\ R, 27 Oct 1998, available at https://www.wto.org/english/tratop_e/dispu_ e/76r.pdf (last accessed 29 February 2016).

Shrimp/Turtle Case[116]

The case is about import prohibition of shrimp and certain shrimp products. In this case, one country stated higher domestic environmental standards and the other showed lower domestic environmental standards as agreed in their local governmental bodies. The country invoking higher domestic environmental standards could justify trade sanctions against a country with lower environmental standards. The GATT under Article 20 talks about exceptions to trade. The issue was whether exceptions under GATT could be applied in the case of GMO. It was viewed that the exceptions under Article 20 of the GATT cover both living and non-living natural resources, including GMOs. GATT Article 20 has been at the center of the most debated environment-related trade disputes. The striking point is the condition that any measures taken to protect human health and the environment must not imply "arbitrary or unjustified discrimination between countries where the same conditions prevail or there is a disguised restriction on international trade." The appellate body held that GATT exceptions under Article 20 must be read in the light of contemporary concerns of the community of nations about the protection and conservation of the environment. The case evolves the approaches that there could be exceptions to the trade in GMOs in the light of local community concerns and the safety of the environment. The appellate body encouraged countries to seek an internationally agreed solution for the preservation of globally shared environmental resources, for instance in the form of a multilateral environmental agreement.[117] Adoption of a consultative approach to resolve conflicts in case of trade in GMO was recommended. The panel and the appellate body in this case felt that the US did not take the desirable or required steps of entering into multilateral environmental agreements to protect turtles. This case presents an open debate between trade and environment. Such situations will have to address the dispute between environment and growth as well. Environmental circles questioned whether international environmental agreements including trade regulatory measures would be respected in a conflict between trade and environment.

---

[116] Appellate Body Report, Doc. WT/DS58/AB/R of 12 October 1998.

[117] Ricardo Melendez and Sanchez. 2007. *Trading in Genes*, p. 71.

Monarch Butterfly Case[118]

In the present case, the release of GMOs into the environment was considered to be risky. It was viewed that there would be a negative impact on the environment from such releases. It was argued that the lacewing, which is the food of butterflies, has been eating GM pollen and disappearance of the monarch butterflies was linked to it.[119] Therefore, extreme caution and care is required while handling the GMOs, in particular before their release into the environment, and every possible step should be taken to negate the adverse impact. In this case, even before the Cartagena Protocol on Biosafety came into being, the precautionary approach was prescribed and recommended for the regulation of GMO release into the environment. Decisions such as this one have probably directly or indirectly influenced the making of biosafety protocol. Scholars viewed the decision in different ways. This approach was considered relevant and appreciated by scholars as it showed negative impact[120] from the deliberate release of GMOs into the environment. The same was criticized to be extreme and negative by international trade community.[121]

Sothern Bluefin Tuna Case[122]

This case involved the trade dispute between New Zealand and Japan as well as Australia and Japan. In this case, specific reference to the precautionary approach of the Cartagena Protocol was made. Prior informed consent procedure was insisted by the importing country from

[118] See Naidu. 2009. *Biotechnology and Nanotechnology*, p. 18.

[119] Sahai, Suman. 2014. 'We do not have the competence to play around with GM foods', *The Times of India* (Kolkata edn), 2 November, available at http://timesofindia.indiatimes.com/home/sunday-times/all-that-matters/We-do-not-have-the-competence-to-play-around-with-GM-foods-Suman-Sahai/articleshow/45009714.cms (last accessed 3 November 2014).

[120] Lossey, John E. 1999. 'Transgenic Pollen harms Monarch Larvae', *Nature*, 399: 214.

[121] See Beringer, John E. 1999. 'Cautionary Tale on Safety of GM Crops', *Nature*, 399: 405.

[122] Sothern Bluefin Tuna case (1999 38, ILM, 1624-1656). See also http://www.ejil.org/pdfs/11/4/555.pdf (last accessed 26 February 2016).

the exporting country. The probable risk that GMOs may pose on the environment and the human health was considered and suggested for precautionary measures, including that of prior informed consent. It was considered very much necessary to ensure the safety of handling, use, and trade of GMOs. At the time of adjudication of this case, the Cartagena Protocol on Biosafety was in the making and approval was sought for certain important principles under the protocol including that of prior informed consent and the precautionary measures.

Greenpeace France[123]

Despite of lack of relevant international judicial decisions, the ECJ attempted to resolve a biotechnology- and trade-related dispute in 2000. It has been observed that in most of the trade disputes involving GMOs the EU is a party, particularly because of varied approaches between the nations, in particular European countries and other countries. Europe has traditionally been cautious regarding GMOs and was strict on insisting for precautionary principle in GMO trade. During the making of the Cartagena Protocol, this particular principle was responsible for heated debates, as a block of countries led by the US was against the principle.

Korea Beef Case[124]

The appellate body, while considering the measures affecting imports of fresh, chilled, and frozen beef from Korea, attempted to update the balancing test for assessing the fulfilment of the requirement of necessity as included in some of the exceptions of GATT. The case was on the similar issues that were adjudicated in the Shrimp/Turtle Case. The issue was whether it would be possible to take measures to stop and regulate the import of Korean beef using the GATT exceptions. These exceptions perhaps would balance the interests of the member states.

[123] Greenpeace France and others, C-6/99, European Court of Justice Reports (ECR) (2000) 1-1651.

[124] Appellate Body Report, Doc. WT/DS161/AB/R-WT/DS169/AB/R of 11 December 2000.

Invoking the exception provisions should not unusually obstruct the trade interest of other member states. According to the appellate body, one crucial factor in this balancing exercise is the relative importance of the common interests of values that law or regulation is intended to protect.

Tuna Dolphin

The relation between the trade and the environment has been the focal point of discussion in this dispute. Interpretation of Article 20 of the GATT, in the light of SPS and TBT agreements of WTO, was contested. Both GATT and the SPS agreement allow member states to adopt measures to protect human health and the safety of the environment even if they are trade restrictive according to the WTO and GATT standards. It was viewed that there shall not be any unscientific and arbitrary restrictions on international trade involving GMO under the garb of protection of human health and safety of the environment.[125] This decision tried to balance the interest of the GMO trade with the protection of environment and health.

Asbestos Case[126]

The WTO appellate body held that the ban on imports of asbestos and asbestos-containing products was covered under the TBT agreement. With respect to import bans on GM products, the TBT agreement is relevant and trade regulations and technical barriers should not be more trade restrictive than necessary to fulfil legitimate objectives including national security, health, and safety of the environment. It is considered a decision in favor of bans on GM products on legitimate grounds for reasonable objectives well within the framework of the WTO and TBT agreements.

---

[125] See Ricardo Melendez and Sanchez. 2007. *Trading in Genes*, p. 180.

[126] *Canada* v. *European Communities* WT/DS135/AB/R R 12 March 2001, available at https://www.wto.org/english/tratop_e/dispu_e/135abr_e.pdf (last accessed 29 February 2016).

MOX Plant Case[127]

Ireland and the UK were the parties involved in the dispute, wherein the requirement of precautionary approach and the prior informed consent were considered positively by the adjudicating body. The case promotes the cause of the Cartagena Protocol regarding safety in GMO trade. The Law of the Sea Tribunal formally recognized the principle of prior informed consent and advocated for precautionary approach in relation to the protection of environment as envisaged under the protocol.

Netherland v. Parliament and Council[128]

It was a case concerning the demand by the Netherlands for an action for annulment of EU directive on the legal protection of biotechnology inventions. Although the ECJ did not subscribe to the argument of the Netherlands, it agreed to review the directive in the light of the CBD.[129] The Netherlands was of the opinion that the EU directive had not considered the obligations and concerns of CBD. The convention attempts to ensure sovereign rights over biological and genetic resources. It provides for advance informed consent, benefit sharing, and acknowledgment of sources in case of the use of biological resources and, to some extent, runs in contrast with the vision of the TRIPS agreement, which aims to provide for commercial exploitation of innovative use of biological resources. It is quite noteworthy to mention here that in 2001, the EU adopted a special directive on the regulation of deliberate release of GMOs into the environment. The directive on the biotechnology inventions intends to regulate biotechnology research and conferment of intellectual property protection while considering the social, cultural,

---

[127] *Ireland v. United Kingdom*, International Tribunal for Law of the Sea (ITLOS) reports, 2001, pp. 95–149, available at https://www.itlos.org/cases/list-of-cases/case-no-10/#c667 (last accessed 27 February 2016).

[128] *Kingdom of the Netherlands v. European Parliament & Council of the European Union*. Case C-377/98. 2001 ECR I-7079, available at http://www.jstor.org/stable/3070691?seq=1#page_scan_tab_contents (last accessed 29 February 2016).

[129] Francioni and Scovazzi (eds). 2006. *Biotechnology and International Law*, p. 38.

and ethical concerns. The directive on the deliberate release of GMOs into the environment intends to ensure safety of the environment during GMO release. The status and validity of the same was challenged by the Netherlands before the ECJ. However, the ECJ found that the directive is well-founded and is not against any international legal measures including CBD.

Monsanto Italia[130]

This is a case where the provisions of SPS and TBT agreements were interpreted in the light of GMO trade. The cautious steps and adoption of trade-restrictive measures prescribed under the SPS and TBT agreements acknowledge a risk specifically associated with the GM products. In such cases, it becomes very important to ascertain the risk with sufficient proof. While interpreting and analyzing the term "risk assessment," the same court ascertained that the existence of specific and identified risks involved in the GMO trade was sufficient to adopt measures against the import of GM products. The court lowered the standard of proof in this case by interpreting the expression "detailed grounds" as requiring mere evidence that indicates the existence of a specific risk.

Republic of Austria Case[131]

The WTO agreements including GATT/TBT/SPS and their effect on GMO trade of member nations was discussed in this case. The Republic of Austria sought the annulment of rejection of draft legislation aimed at establishing a GMO-free area in Upper Austria by commission, before the ECJ. Earlier on 2 September 2003, the European Commission (EC) had rejected the proposal of Republic of Austria for declaring it a GMO-free area. It is worth noting here that in 1998 the EU adopted a

---

[130] *Monsanto Agricoltura Italia Spa and Others v. Presidenza del Consiglio dei Ministri and Others* Case C-236/01, ECR-I, 8105, available at http://curia.europa.eu/juris/showPdf.jsf?docid=71432&doclang=en (last accessed 29 February 2016).

[131] *Republic of Austria v. European Commission* T-366/03 and T-235/04 (2004) OJC/21/20, available at http://eur-lex.europa.eu/legal-content/EN/TXT/?uri=CELEX%3A62005CJ0439 (last accessed 29 February 2016).

directive on the legal protection of biotechnology inventions. The directive is operational across the Europe, including the Republic of Austria, encouraging and protecting biotechnological inventions. As a matter of concern, while promoting biotechnological inventions practically, trade in GMOs cannot be banned. The TRIPS agreement also encourages biotechnology patents as part of broader WTO/GATT framework. In fact, WTO nations are obligated to protect intellectual property rights in biotechnology inventions and promote trade in biotechnology and GMOs. The EU, which earlier was not open to GMO trade and had imposed moratorium on GMO, lifted the moratorium afterwards due to the international pressure. Probably all these reasons have influenced the EC to reject the proposal of Republic of Austria to declare Austria as GMO-free trade area.

Japan Apple Case[132]

The case presents the interface between the WTO agreements and the Cartagena Protocol on Biosafety. In the background of the Cartagena Protocol on Biosafety, it was viewed that nations can look beyond the WTO framework, in particular the SPS agreement, to ensure safety of the environment by adopting precautionary approach as enshrined under the protocol. The WTO appellate body has pointed out the failure on the part of Japan to invoke precaution as a principle separate and distinct from the provisions of the SPS agreement.[133] At the same time, the appellate body indirectly recommended the approach of the protocol with reference to the trade issues involving GMOs. It was decided that in cases of GMO trade if the WTO agreement framework cannot find sufficient solution, the biosafety protocol should be invoked to ensure the safety of the environment as the protocol prescribes stricter threshold. WTO agreements and the protocol could be read together to deal with the GMO trade. It is worth noting here that the risk assessment purpose under the SPS agreement is the precautionary approach meant for ensuring safety of the health and the environment. At the same time, precautionary approach under the protocol is also meant for ensuring the safety and security of

---

[132] Appellate Body Report Doc. WT/DS245/AB/R of 26 No 2003.

[133] Francioni and Scovazzi (eds). 2006. *Biotechnology and International Law*, p. 42.

the environment. From this perspective, both the SPS agreement and the protocol have definite interfaces and mutually serving objectives.

Monsanto Agricultura Italia

The EPJ was asked to resolve a trade dispute involving GMOs in agriculture. This case was decided at the time when the Cartagena Protocol on Biosafety had already been adopted, recommending precautionary principle, advanced informed consent, and such other safety measures for the release of GMOs into the environment and for their trade. As already established in the previously discussed trade disputes since 1997, the EU is quite sensitive about the potential risks posed by the GMOs and their release into the environment. By the time of this decision, principles of caution and precaution were clearly established in the EU. The present case reiterated and further established the precautionary principle and avoiding such other risk and security measures for the regulation of trade in GMOs.

Eva Clawischnig[134]

Traditionally, the EU has always been hesitant and cautious regarding the issue of GM products and crops, it is quite expected from the ECJ to echo the language of the European traditional views and perspectives. However, at the same time it needs to be noted that the intention was not meant to put restrictive measures to hurt trade interests but to take care of the safety of the environment. As always, the ECJ held that the commercialization of GM products should not result in damage to the environment and the human health. It is trade versus environment, and the logical conclusion should echo the language of promoting trade only after ensuring safety of the environment.

Ministero della Salute v. Codacons et al.[135]

Further in 2005, the ECJ once again adopted a cautious approach toward GM products. Since GM products are associated with potential risks

---

[134] Eva Glawischnig C-316/01, [2003] ECR 1-5995.
[135] *Ministero della Salute v. Codacons et al.* C- 132/03, 2005, ECR: I, 4167.

that could harm health and environmental safety, the ECJ was cautious in this regard. It is interesting to note here that by this decision of ECJ, the Cartagena Protocol on Biosafety came into picture, which adopted the principle of precaution, advance information, and other such measures to ensure safety of the environment and the health of the living beings. These principles have been the focal points in number of decisions of the ECJ while deciding on the disputes involving GM products.

Upper Austria Case[136]

Upper Austria had prepared a draft legislation for establishing Upper Austria as a GMO-free area. Earlier on 2 September 2003, the EC rejected the proposal of Upper Austria for declaring it as a GMO-free area. The region of Upper Austria appealed against and sought the annulment of a decision taken by EC. It was stated that in the light of WTO/GATT obligations to which the EU is a party, declaring Upper Austria as a GMO-free area is not feasible and such legislations aiming GMO-free area cannot be allowed.

Pfizer Case[137]

Precautionary principle is perhaps well-entrenched in the EU law and has been given a robust interpretation in the European judicial practice.[138] In the present case, it was stated that precautionary approach was necessary and should be adopted before releasing the GMOs into the environment. The precautionary principle is creating binding obligations with relation to biotechnology trade and development. It is presumed that the adoption

---

[136] *Upper Austria v. European Commission* (T-366/03 and T-235/04) (2004) OJC/35/11, available at http://eur-lex.europa.eu/legal-content/EN/TXT/?uri=CELEX%3A62005CJ0439 (last accessed 29 February 2016).

[137] *Commission of the European Communities v. CEVA Sante Animale SA and Pfizer Enterprises* 2005 ECJ CELEX LEXIS. 737, available at http://curia.europa.eu/juris/document/document.jsf?text=&docid=60407&pageIndex=0&doclang=EN&mode=lst&dir=&occ=first&part=1&cid=123185 (last accessed 29 February 2016).

[138] Francioni and Scovazzi (eds). 2006. *Biotechnology and International Law*, p. 19.

of the precautionary approach would play a vital role with reference to the trade in GMO and the relations between the nations involved in the biotechnology trade. It is perceived that the precautionary approach is part and parcel of every state's due diligence obligations to ensure that biotechnology-related activities are properly regulated and kept under control to prevent environmental damage and adverse effects on health as per the language of the SPS agreement and the Cartagena Protocol.

Commission v. Italy[139]

The dispute involves infringement proceedings against the non-compliant member regarding trade in biotechnology. As an aftermath of the adoption of the EU directive in 2001 on the deliberate release of GMOs into the environment, uniform and common rules were prescribed across Europe. The EC is not only responsible for the implementation of the directive but also takes note of any non-compliance by the member states. It is a well-established fact that the EU prescribed a more strict and cautious approach toward GMO trade. Under the directive, strict regulation was imposed on the GMO release into the environment. Probably a few European nations had a slightly different approach in this regard, which was brought to the attention of the EC for proper directions to be issued to the member states regarding proper compliances of the said directive.

EC Biotech Case[140]

The EC biotech case discussed the role of international law in the interpretation of WTO rules[141] before the WTO dispute settlement panel.

---

[139] *Commission of the European Communities v. Italian Republic* (Case C-456/ 03) - [2005] All ER (D), available at http://curia.europa.eu/juris/document/ document.jsf?text=&docid=59370&pageIndex=0&doclang=EN&mode=lst &dir=&occ=first&part=1&cid=192263 (last accessed 29 February 2016).

[140] European Communities—Measures Affecting the Approval and Marketing of Biotech Products, WTO Dispute Settlement Body decision: WT/ DS291/R (US), WT/DS292/R (Canada), WT/DS293/R (Argentina), 29 September 2006.

[141] Wuger and Cottier (eds). 2008. *Genetic Engineering and the World Trade System*, p. 171.

In particular, the relationship between the Cartagena Protocol and the WTO–SPS agreement was discussed.[142] The EC–Biotech case was initiated in May 2003 on the request of the US, Canada, and Argentina for consultations on certain measures taken by the EC. In 2006, the European Council took a decision to impose moratorium on GMO approval and marketing of biotechnological products. But it was alleged that the European communities had maintained a moratorium on the approval of new agricultural biotechnology products since October 1998, and due to this moratorium, all product applications in the system had been stalled indefinitely, giving rise to de facto product-specific market bans.[143] An important point to note was that not a single biotechnological application under consideration during this time had been approved. The de facto moratorium on the commercialization of GMOs in the EC, which existed from 1998 until May 2004, arose, on the one hand, from the reluctance of individual EC member states to allow the commercialization of GMOs before the existing regulatory framework had been enhanced, and on the other hand, from the need of member states to respond to internal public and consumer concerns regarding GM technologies and products.[144] The parties to the complaint had not challenged the procedures of approval of biotechnology products, but rather the application of these procedures. According to one of the directives, an EU member state could provisionally restrict or prohibit the use or sale of a biotechnology product in its territory if it has justifiable reasons to consider that a product constitutes a risk to human health or the environment. The EU contended that the general timeframe specified for decisions in the Cartagena Protocol on Biosafety was 270 days. Apart from this, in Article 8 of the protocol the member states had the right to request additional information and, accordingly, extend the timeframe for decision-making. In addition, Article 12(1) of the protocol affirmed on the parties the right to review and change decisions on imports of new scientific information on potential adverse effects on biodiversity. The EC argued that in each of the cases the delay

[142] Hocking. 2009. *The Nexus of Law and Biology*, p. 100.

[143] Howse, Robert L. and Henrik Horn. 2009. *European Communities— Measure Affecting the Approval and Marketing of Biotech Products*, World Trade Review, 8(1): 49–83.

[144] Somsen. 2007. *Regulatory Challenge of Biotechnology*, p. 112.

could be justified on the basis for additional information. It was further argued that even though the EC regulator had concluded that certain products were safe, other scientists had raised legitimate questions about the adequacy of the risk assessment. Based on the precautionary principle, it was not unreasonable for the member states to provisionally ban the products in question.

Panel Report

The panel reached the conclusion to use a non-WTO multilateral agreement, such as the Biosafety Protocol, as a tool to interpret WTO agreements if all WTO members are also parties to that multilateral agreement. Given the unlikelihood of such a correspondence with the membership of multilateral conventions, any use of international law in the interpretation of WTO provisions would occur rarely, if at all, in some experts' opinion cutting-off those provisions from the rest of the international law.[145] The panel was of the view that in the case at hand, the rules of international law were not applicable in relations among all the parties to the dispute. The panel did not take a position on whether, in such a situation, it would be entitled to take into account the other relevant rules of international law.[146] It is pertinent to note here that several WTO members, including the US and other parties to the dispute, were not parties to the Biosafety Protocol. Probably due to this reason, the panel thought that it was not required to take into account the Biosafety Protocol provisions in interpreting the WTO agreements at issue. It was found that in the specific case, Biosafety Protocol, despite its high number of ratifications with as many as 139 countries following it, could not be used as an interpretative tool for WTO agreements. The panel concluded that the general and product-specific moratoria had led to an "undue delay" in the completion of the EU's approval procedures for biotechnology products, thus breaching the EU's obligations under

---

[145] Study Group of the International Law Commission Report. 'Fragmentation of International Law: Difficulties Arising from the Diversification and Expansion of International Law,' International Law Commission, fifty-eighth session, A/CN.4/L.68213, April 2006.

[146] Wuger and Cottier (eds). 2008. *Genetic Engineering and the World Trade System*, p. 172.

the SPS agreement. With regard to the issue of member states prohibiting or restricting biotechnology products, the panel held that sufficient scientific evidence was available to carry out a risk assessment. The panel therefore rejected the EU's defense of the bans as precautionary measures under Article 5.7 of the SPS agreement. The report called on the EU to bring the measures in conformity with the SPS agreement, which would imply revoking them or providing an SPS agreement-compliant risk assessment to justify the measures. Although, the panel report has said that the measures taken by EC were not justified under various international agreements, it did not provide any sanction against the EC. The panel has said that the EC should bring its measures within the ambit of the international agreements and suggested that the two parties sit together in order to find a solution. However, the panel itself has not provided any solution to the dispute. It is also seen that the present dispute will be a very difficult dispute to settle. This is because of the vast cultural and political differences between the two parties. Such differences cannot be put aside so easily.

Trade and Environment: Can Both be Balanced in the Era of Biotechnology

The central theme of the Cartagena Protocol being the regulation of transboundary movement of LMOs, naturally other international agreements related to trade need to have some manner of logical interaction with it. The other agreements in question include the Law of the Sea, international transit and transportation arrangements, and international health agreements that address human pharmaceuticals. However, the body of law that was on everyone's minds was the multilateral system of trade rules embodied in the WTO. Article 20 of the GATT gives countries the right to adopt measures for protection of the health of their populace or biodiversity, as long as the measures are not discriminatory or protectionist in nature. The SPS agreement specifically addresses food safety, and animal and plant health standards. The decision-making procedures of the current international regulatory structure for GMOs, in the form of SPS and the Protocol,[147] share similar elements and are

---

[147] See generally Evenson and Raney (eds). 2007. *The Political Economy of Genetically Modified Foods*, pp. 293–9.

compatible in a significant manner.[148] Decisions under both agreements must be based on risk assessment, that is, scientific approach to the identification and evaluation of the risks of imports, ensuring informed and objective choices. Moreover, the results of such assessments are not determinative in nature; they must be integrated with socioeconomic considerations to determine the actual risk in a contextual application. Finally, in the absence of sufficient evidence, both the instruments enable the governments to make decisions regarding the introduction of GMOs, based on precautionary principle. Last but not least, under the TBT agreement, measures such as labeling, symbols, and health and environmental standards should not act as obstacles to foreign competitors. This makes the labeling of LMOs under the Cartagena Protocol a matter of concern, with countries trying to develop their biosafety regimes being required to consider the possible implications of including labeling requirements for LMOs with respect to the TBT. The WTO rules are liked by those whose major concern is protectionism masquerading as environmental protection. Although Article 34 of the protocol does make the parties commit to establish mechanisms to address non-compliance, among other things, to avoid the conflict from reaching the WTO, though wherever serious commercial interests are at stake, the WTO will probably be the forum for settlement of disputes. And on such occasions it would be very difficult for the dispute panel to ignore the strong wording of the protocol[149] as it is "non-subordinate" to other international instruments. It is observed that the protocol goes beyond the mechanism of SPS agreement in terms of ensuring safety and health of the living beings. The significance of the protocol's precautionary provisions assumes greater prominence when they seek to fill in certain gaps in the SPS agreement. Let us examine how exactly the protocol goes with and, at the same time, goes ahead of SPS agreement:

1. The SPS agreement does not spell out exactly what a risk assessment entails, but the protocol does so in detail in Annex II.

---

[148] Wuger and Cottier (eds). 2008. *Genetic Engineering and the World Trade System*, pp. 208–9.
[149] See generally Evenson and Raney (eds). 2007. The *Political Economy of Genetically Modified Foods*, pp. 293–9.

2. The SPS agreement does not mention risk management, but only risk assessment, whereas Articles 15 and 16 of the protocol make it clear that both exercises are necessary, defining the latter as the gathering of the data, and the former as the building of a regulatory regime based on that data.[150]

3. The protocol explicitly allows parties to take into account socio-economic considerations while taking decisions, whereas the SPS agreement says nothing on the subject.

4. The protocol is specific about the process for review of decisions in the light of new evidence,[151] whereas the SPS agreement is ambiguous about how to treat measures adopted provisionally in the face of uncertainty.

5. The provisions in Article 15 of the protocol go some distance toward laying the onus on the exporter to establish the harmless nature of the LMO in question.[152] The SPS agreement, on the other hand, is silent on this issue.[153]

The provisions of the protocol therefore enrich the SPS agreement by adding details that help in the operationalization of the precautionary principle in the context of LMOs. This lends support to the argument that trade and environment agreements should be mutually supportive. Moreover, the protocol also arguably establishes the precautionary principle[154] as a principle of international environmental law, and since it can be used for protection of human health and socioeconomic considerations of customary international law.

---

[150] It further sets out some guidance in creating the regime such as asking parties to try to ensure that any LMO should undergo an appropriate period of observation commensurate with its life cycle or generation time before it is put to its intended use.

[151] Article 12 of the Cartagena Protocol.

[152] Paragraphs 2 and 3 of Article 15 state that the party of import may require the exporter to carry out the risk assessment, and the notifying party to foot the bill.

[153] Cosbey and Burgiel. 2000. 'The Cartagena Protocol on Biosafety: An Analysis of Results'.

[154] See generally Evenson and Raney (eds). 2007. *The Political Economy of Genetically Modified Foods*, pp. 293–9.

India's Stand on Biotechnology Trade and Environment

Trade in biotechnology in India is new idea when compared to the US and the European countries. However, it is not that India has not made any progress in the field of biotechnology. The Indian biotechnology sector has taken shape through a number of scattered and sporadic academic and industrial initiatives.[155] India needs to take biotechnology seriously, as its agriculture sector is facing a daunting task of having to produce more farm commodities for the ever-increasing population. Biotechnology would help to win this challenge. However, the Government of India and its various agencies have to address the social concerns of the people of the country in order to propel the emergence of biotechnology in India. For this, the government and the industry have to work together to advance the benefits of modern biotechnology while, at the same time, address the concerns and interests of the public. Apart from addressing public concerns, the government has to address the issue of lack of coordination among several agencies that play a role in the Indian biotechnological regulatory framework.[156] It has been stated that in dealing with several agencies, companies experience an approval process that causes significant confusion and delays in commercialization,[157] as manufacturers are required to acquire approvals from multiple state, district, and federal agencies for routine activities.[158] Therefore, it is imperative that agencies dealing with biotechnology adopt procedures that are not tedious or complex. Apart from procedural problems, it is claimed that that there is a lack of expertise regarding biological sources

[155] National Biotechnology Development Strategy Draft, Department of Biotechnology, Ministry of Science & Technology, Government of India.

[156] European Business Technology Center. 2010. 'Biotechnology in India: Its Policy and Normative Framework', pp. 3–130. New Delhi: European Business Technology Center. Available at http://ebtc.eu/images/ebtc-website-assets/publications/sector-publications/biotechnology/market-reports/101010_REP_biotechnology-in-india_its-policy-and-normative-framework.pdf (last accessed 16 July 2016).

[157] Frew, S.E. 2007. 'India's Health Biotech Sector at Crossroads', *Nature Biotechnology*, 25.

[158] Lager, Erics. 2008. 'Biologics Regulation in India', *BioPharm International*, 21(3), available at http://www.biopharminternational.com/biologics-regulation-india-0 (last accessed 16 July 2016).

on the part of some regulatory agencies, while others have pointed out staffing problems. Biotechnology must evolve as an environmentally sustainable science and must not introduce anything harmful to the environment, so there is a need for proper health risk assessment. In terms of biosafety, India has a strong regulatory system; however, the existing system now requires to ensure that all GM crops undergo rigorous reviews and safety assessments prior to their import, field-testing, or release. According to scholars Suman Sahai,[159] Vandana Shiva, and Jayashree Watal, countries like India do not have proper and established regulatory mechanism to ensure safety of the biotechnological crops and foods. There is lack of an effective implementation mechanism. It is also urged that the penalties prescribed under various laws must be significantly enhanced to create adequate deterrence. A biosafety regulation specifically for agro-food biotechnology products is the need of the hour. The proposed Biotechnology Regulatory Authority should be established at the earliest to deal with trade and environment issues in biotechnology. It can comprise members from academics, social sciences, government, scientific community, economics, industry, and so forth, and the authority can take a case-by-case approach in deciding merits of activities and events.[160]

The Biological Diversity Act, 2002

India, being a party to the CBD signed at Rio-de Janerio on 5 June 1992 enacted the Biological Diversity Act, 2002. The basic objectives of the act are conservation of biological diversity, sustainable use of its components, and fair and equitable sharing of the benefits. One of the objectives of the act, that is, sharing of the benefits made by the use of genetic resources, is defined to include inter alia the appropriate transfer of relevant technology, taking into account all rights to technologies. In India, the Biological Diversity Act, 2002 in a cursory manner regulates the area of operation for GMOs. It is felt that there are serious concerns about the fact that humans are meddling with nature and this exercise

[159] See Sahai. 2014. 'We do not have the competence to play around with GM foods'.

[160] Dhawan, Vibha. 2004. *Bt for Food and Nutritional* Security. TERI.

may have detrimental effects on biodiversity, environment, ecosystem, health of humans and animals, and so forth. The genetic material of plants, animals, and microorganisms has been considered a part of "biological resources" under the act. It is pertinent to note that these concerns addressed in the Biological Diversity Act, 2002 are not holistic in their approach over concerns in trade-related aspects of biotechnology and its impact. The law broadly prohibits any foreign person who is not a resident of India or foreign corporation that is not incorporated in India from "obtaining any biological resource[161] occurring in India or knowledge associated thereto" for trade or commercial utilization[162] without the prior approval of the National Biodiversity Authority. Regulation of bioprospecting by Indian resident citizens and Indian corporations for trade purposes is left to state biodiversity boards.[163] The law, in turn, requires the National Biodiversity Authority to secure equitable benefit-sharing for the use of "accessed biological resources, trade in their products and by-products ... and knowledge relating thereto."

Future of Biotechnology Trade and Development: The Way Forward

It is necessary to gauge that for any nation a chief reason that adds to the burden of complying with any multilateral legal regime on GMOs is the dilemma of what the law is. There is clearly no comprehensive regime dealing with this issue either at the international or at the national front. Further, the law is subject to various interpretations; hence, it is believed that the existing law is not unambiguous and lucid and thereby enables

[161] Section 3(1). "Biological resource" broadly encompasses "plants, animals and microorganisms or parts thereof, their genetic material and by-products ... with actual or potential use or value"; Section 2(c). Biological resources do not include "products which may contain portions or extracts of plants and animals in unrecognizable and physically inseparable form.

[162] Section 3(1). "Commercial utilization" refers to end uses of biological resources for commercial purposes, such as for drugs, food flavours, cosmetics, colours, and "genes used for improving crops and livestock through genetic intervention."

[163] Sreenivasulu, N.S. and Arnab Sengupata. 2010. 'Biological Resources, IPR and Biodiversity', Manupatra Intellectual Property Reports, 1(4): F35–F47.

nations to escape the same by advancing a convenient line of reasoning that falls within one of the loopholes. In addition, if the laws are not clear, then sustainable development, which is the undercurrent of the entire GMOs issue, will not become a reality. Sustainable development is a new name of economic envisaging a process of change, in which economic and fiscal policies, trade and foreign policies, energy, and agricultural and industrial policies all aim to induce development paths that are economically, socially, and ecologically sustainable. The late twentieth century has seen a burgeoning and widespread international attention on issues pertaining to the technological advancement. In the midst of this debate, "sustainable development" is the most elusive concept that has originated.[164] On one hand, as a societal construct, sustainable development demands that people recognize themselves as concurrently trustees of today's resources and guarantors of tomorrow's, but on the other hand, as a legal paradigm, sustainable development is bedevilling and ubiquitous. Although the concept is not an evolution of a particular law and only a paradigmatic shift, it has taken different shades of meaning as the various environmental, political, social, and economic groups have struggled to define it and produce practical results in the real world. It is essential to understand that the entire debate on GMOs trade and biosafety is grounded in the recognition of the urgent need to ensure sustainable development. The Supreme Court of India in *Vellore Citizens Welfare Forum v. Union of India*[165] recognized the development of the concept of sustainable development and its complex pedigree. The Cartagena Protocol on Biosafety,[166] which India has ratified as one of the eight principles of sustainable development, is the precautionary principle permeating the ethos of the document. Hence, the protocol endeavors to achieve sustained development and what is required on the part of the nations is a strong and effectively worded domestic legislation that is guided by the national policies of the country. The WTO agreements and the Cartagena Protocol pursue different goals. While

---

[164] See Wuger and Cottier (eds). 2008. *Genetic Engineering and the World Trade System*, p. 5.

[165] AIR 1996 SC 2715.

[166] See generally Evenson and Raney (eds). 2007. *The Political Economy of Genetically Modified Foods*, pp. 293–9.

the WTO mandates non-discrimination in international trade between member states, the protocol regulates the trans-boundary movement of LMOs produced by modern biotechnology. Some potential areas of conflict, linkage, and accommodation between the two regimes include risk assessment for the imposition of measures permitted under each regime, the degree of scientific evidence required, the underlying objectives, the burden of proof, and dispute resolution.[167] With respect to Indian position in this matter, it is yet to be seen as to what extent biotechnology is promoted and implemented in India. With the recent news that GMO crops have been given the green signal by the court, one can say that biotechnology in India is moving forward. However, whether states in India will support biotechnology products remains to be seen.

[167] Kelly, Claire R. 2006. 'Power, Linkage and Accommodation: The WTO as an International Actor and Its Influence on Other Actors and Regimes', *Berkeley Journal of International Law*, 24: 79.

6   Human Rights Concerns in Biotechnology

U ndoubtedly, biotechnology has opened up many new horizons in the scientific world. However, the scientific progress made in this field is also accompanied by many problematic questions from both legal and human rights point of views. While a human rights approach is only one of the ways to deal with the problem, it has the advantage of combining law and ethics.[1] An inquiry into the integration of law, human rights, and ethics reveals the existence of the fundamental principles of natural law even before the formulation of formal moral law and morality.[2] At the same time, the fundamental notions of human rights are also universal in nature and character. A jurisprudential enquiry into the concept of life reveals that there are few intrinsic values attached to life. The conceptual framework of life is connected to the

[1] Francioni, Francesco and Tullio S. Covazzi (eds). 2006. *Biotechnology and International Law*, p. 369. Portland, USA: Hart Publishing.

[2] See generally, Chrost, A.H. 1974. *An Introduction to Aquinas*, 19 A.M.J. of Jurisprudence 1.

natural law that advocates for the inherent values of life such as dignity, integrity, sustenance,[3] survival,[4] and self-preservation.[5] These values are often being referred to as human rights by the society.

Nature has provided every living being a right to self-dignity and integrity. Every living being, be it a plant, animal, or a microorganism, deserves a drive for self-preservation of natural features attributed to it by the nature. Every living being has a right to preserve the intrinsic values of life, which should not be tampered.[6] But biotechnology is capable of changing the natural features and incorporating certain novel features into living beings, thus entering into the realm of infringement of human rights. Such manipulation of living beings hits at the inherent dignity, integrity, and natural setup of living beings and nature itself. Furthermore, it disturbs the sustenance and self-preservation of natural features of life. Living beings are part of nature. Alteration or manipulations of any living being like plants, animals, or microorganisms strike at the integrity and balance of the nature and human beings.[7] Thus, it has been argued that certain scientific and technological[8] developments, such as biotechnology, have the tendency to disturb and violate the intrinsic values of life and its rights.[9] Such technological interventions of biotechnology in the intrinsic structure and values of life and living beings ignite debates on the prospects of biotechnology from the perspective of human rights and philosophy. The various international conventions and the domestic legal frameworks of the United States (US), the European Union (EU),

---

[3] Freeman, M.D.A. (ed.). 1994. *Lloyd's Introduction to Jurisprudence*, 6th edn, p. 81. London: Sweet and Maxwell.

[4] See generally, Hart, H.L.A. 1992. *The Concept of Law*, p. 188. London: Oxford University Press.

[5] Hart. 1992. *The Concept of Law*, p. 188.

[6] Sreenivasulu, N.S. 2008. *Biotechnology and Patent Law: Patenting Living Beings*, p. 196. Noida: Manupatra Publications.

[7] See generally, Unger, Roberto, *Knowledge and Politics*, as cited in M.D.A. Freeman (ed.). 1994. *Lloyd's Introduction to Jurisprudence*, 6th edn, pp. 595–7. UK: Sweet and Maxwell.

[8] See generally, Conant, James B. 1953. *Modern Science and Modern Man*, pp. 97–8. New York: Columbia University Press.

[9] Radhakrishnan, Sarvepalli and P.T. Raju (eds). 1995. *The Concept of Man: A Study in Comparative Philosophy*, p. 17. Harper Collins.

and India on human rights directly or indirectly provide for the protection and preservation of human values, and it is interesting and even essential to conduct an inquiry into the fundamental notions of these rights in the wake of developments in biotechnology.

International Conventions on Human Rights and Biotechnology

Research in biotechnology, particularly in human genetics, has evolved, opening a complexity of balancing human rights issues with the promotion of research and its developmental results. The primacy of human dignity and human rights in the context of biogenetics must first be achieved by means of strict application of existing human rights standards as recognized by the international law. There is a range of human rights–related conventions that would have a monitoring effect on the research and development (R&D) in biotechnology.

The first human rights–oriented international convention that comes into picture is the Universal Declaration of Human Rights (UDHR). The declaration is not only relevant for the awareness and protection of general human rights, but is also significant for any scientific development or technological advancement that would involve human beings and their lives. Therefore, the basic canons of the human rights declaration, including dignity, integrity, and self-sustenance, are equally applicable to research in biotechnology that involves the human element. In fact, it is a first attempt at an international level to map various rights, liberties, and freedoms for human beings.

The International Covenant on Civil and Political Rights (ICCPR), an exclusive framework promoting civil and political rights of the civilians, contemplates that no one shall be subjected to medical or scientific experimentation without a free consent.[10] In the context of biotechnology and genetic engineering research, the standard rule of the convention applies in terms of prescribing free consent for biomedical or biogenetic research. The International Covenant on Economic, Social and Cultural Rights (ICESCR) is another convention promoting economic and

---

[10] Francioni, F. and T. Scovazzi (eds). 2006. *Biotechnology and International Law*, p. 326. Oxford and Portland, Oregon: Hart Publishing.

cultural rights of the people at large. These two conventions postulate respect for human rights with special focus on civil, political, economic, and cultural rights.[11] The civil and political rights may not have a direct impact on R&D in biotechnology but economic and cultural rights do.

The rights obtained on the use and exploitation of biotechnology inventions and the trade in turn give rise to commercial and economic rights. At the same time, traditional and tribal people who have conserved and preserved biological and genetic resource as a part of their lifestyle and culture would be affected if the same resources are used for biotechnological innovations and development. These rights are being directly or indirectly affected by R&D in biotechnology. Meanwhile, the EU has been proactive in the field of human rights promotion and related activities. There are various European regional conventions related to human rights in general as well as affecting biotechnology and biomedical research.[12] The EU Convention for the Protection of Human Rights and Fundamental Freedoms is quite significant in this regard. The convention standing in line with the UDHR and other such human rights conventions postulates the inherent dignity of human beings. Inventions in violation of human dignity and freedom are considered violative of human rights under the convention. Particularly in the times when various inventions involve genetic material including DNA, genes, and cell lines, the convention attempts to balance the progress of research with human dignity.

The EU adopted another convention for the protection of human rights and dignity of the human beings related to the application of biology and medicine, the Convention on Human Rights and Biomedicine, in 1997. The convention is also open to states that are not members of the EU.[13] The convention says that research in human beings, tissues, organs, or human genome[14] should be undertaken only after obtaining

[11] See generally, Sreenivasulu, N.S. 2008. *Human Rights: Many Sides to a Coin*. New Delhi: Regal Publications.

[12] See Plomer, Aurora and Paul Torremans. 2009. *Embryonic Stem Cell Patents, European Law and Ethics*, p. 212. Oxford: Oxford University Press.

[13] Francioni and Scovazzi (eds). 2006. *Biotechnology and International Law*, p. 293.

[14] Article 13: 'Interventions on the Human Genome' of the Convention on Human Rights and Biomedicine.

informed consent of the person concerned after disclosing possible associated risks.[15] The interest and welfare of the human beings will prevail over the sole interest of science.[16] At a time when biomedical research on human beings is flourishing, the convention aims to bring in some standards to be followed in terms of respecting human dignity and integrity while embarking on such research initiatives.[17] Using human body and its parts for financial gain is prohibited under the convention.[18] Keeping in mind the recent development in embryonic research, the convention intends to maintain the dignity of the human life by providing adequate protection from the misuse of human embryos. In this connection, creation of human embryos, even if it is solely for research purpose, is considered prohibited[19] under the convention.[20]

The Paris Protocol to the Convention on Human Rights and Biomedicine, adopted in 1998, intends to state that any intervention seeking to create a human being genetically identical to another human being, whether living or dead, is prohibited. Virtually the protocol prohibits human cloning and research in creating genetically modified (GM) or identical human beings.

In 2002, the Strasbourg Protocol to the Convention on Human Rights and Biomedicine was adopted to address the human rights issues involved in biomedical research leading to the development of human tissues and human organs. The protocol intends to ensure human dignity and associated values in developing and transplanting human organs and tissues through biotechnology and genetic engineering.

The Third Protocol concerning Biomedical Research was adopted in 2005, again in Strasbourg. The third protocol is based on the possibly competing assumptions that: (a) progress in medical and biological sciences, in particular biomedical research, contributes to saving lives and improving quality of life, and (b) given the paramount importance

---

[15] Chapter VII of the Convention, Articles 21 and 22: Prohibition of Financial Gain and Disposal of a Part of the Human Being.

[16] Article 5, General Rule, Convention on Human Rights and Biomedicine.

[17] Sreenivasulu, N.S. 2012. 'Patenting of Biotechnology and Human Rights', *International Journal of Law and Policy Review*, 1(2): 146–58.

[18] Article 2: Primacy of the human being.

[19] See Plomer and Torremans. 2009. *Embryonic Stem Cell Patents*, p. 76.

[20] Article 18: Research on Embryo In Vitro.

of the protection of the human beings participating in the research, bio-medical research that is contrary to human dignity and human rights should never be carried out.[21] Sanctity of human dignity means the dignity of the individual person that prevails over both the general inter-est in research and scientific progress and any other interest of society as a whole.[22]

Furthermore, many countries adopted a Universal Declaration on the Human Genome and Human Rights (UDHGHR)[23] to address the issues related to human rights in the development of science and technology, par-ticularly in the field of biology and genetics.[24] While recognizing freedom of research as a part of freedom of thought,[25] the declaration encourages research in the field of biology, genetics, and medicine to guarantee the right to health and relief from sufferings. Acknowledging the amount of research that is being carried out on the human genome, it has been stated under the declaration that research in human genome must respect ethical standards and human dignity. Human genome is considered the heritage of humanity and as such the same cannot be appropriated by any private individual. Any R&D in biotechnology with regard to human genome must be by, for, and from the public interest perspective, that is, for the common good of general public and for serving the purpose of larger inter-ests of the society and the mankind.

The declaration states that the responsibilities inherent in the activities of researchers, including meticulousness, caution, intellectual honesty, and integrity while carrying out their research as well as in the presentation and utilization of their findings, should be the subject of particular attention for the framework of research on the human genome because of their ethical and social implications. Public and

---

[21] See Preamble to the Third Protocol on Biomedical Research to the Convention on Human Rights and Biomedicine.

[22] Article 2 of the Convention on Human Rights and Biomedicine.

[23] United Nations Universal Convention on Human Genome and Human Rights, 1997.

[24] See Vyas, Jaimini, N.S. Sreenivasulu, and Kartikeya Astana. 2016. 'Biotechnology, Genetic Databases and Human Rights', in N.S. Sreenivasulu (ed.), *Human Rights and Development*, pp. 139–82. Bloomington, Indiana, US: Penguin-Partridge Publications.

[25] Sreenivasulu. 2008. *Biotechnology and Patent Law*, p. 208.

private science policymakers also have particular responsibilities in this respect. The policy of the declaration states that research interests in the fields of biology and medicine should not prevail over the respect for human rights and human dignity.[26] In order to protect human rights and fundamental freedoms, limitations to the principles of consent and confidentiality may only be prescribed by law, for compelling reasons within the bounds of public international law and the international law of human rights. The declaration directs the member states to come up with proper legal measures to respect ethics and human rights,[27] and to establish committees for assessing the ethical and human rights issues raised by research on human genome and its applications. The declaration aims that the states should provide for the framework for the free exercise of research on the human genome. At the same time, the declaration sets out that human dignity[28] should not be compromised and any biogenetic-related act that is detrimental to the human values and dignity should be penalized. Genetic heritage, which is part of human dignity and primacy, should be respected and preserved. The declaration emphasizes that research on the human genome should fully respect human dignity, freedom, and human rights, and prohibits all forms of discrimination based on genetic characteristics.

In 1998, the European Council (EC) adopted a protocol on the Prohibition of Cloning of Human Beings. However, the protocol was not very clear on concerns related to human dignity being applicable only to reproductive cloning or if it would also apply to therapeutic cloning. This was solved by the EU Directive on Biotechnology Inventions, which was adopted later in 1998.

Another initiative in this regard, namely, International Declaration on Human Genetic Data, was adopted by the UNESCO in October 2003.[29] The declaration deals with the protection of human genetic data and the issues associated with it, its initiative include ensuring authenticity,

[26] Article 10 of the Declaration of Human Genome and Human Rights.

[27] See Kamstra, Gerald, Mark Doring, Nick Scott Ram, Andrew Sheard, and Henry Wixan. 2002. *Patents on Biotechnological Inventions: The E.U Directive*, p. 210. London: Sweet and Maxwell.

[28] See Plomer and Torremans. 2009. *Embryonic Stem Cell Patents, European Law and Ethics*, p. 211.

[29] Francioni and Scovazzi (eds). *Biotechnology and International Law*, p. 299.

confidentiality, and privacy of the genetic data. Confidentiality of the genetic data is a part of individual's right to privacy and dignified life. It is proposed that breach of confidentiality and abuse of genetic data amount to violation of personal privacy. Furthermore, use of genetic data for therapeutic and progressive purposes also needs to be handled carefully.[30] The perception of UNESCO that human genome is the heritage of humanity is also confirmed by the declaration stating benefits resulting from the use of human genetic data, human proteomic data, or biological samples collected for medical and scientific research should be shared with the society as a whole and also with the international community. The declaration states that when human genetic data or human proteomic data are collected for the purposes of forensic medicine or in civil, criminal, and other legal proceedings, including parentage testing, the collection of biological samples, in vivo or post mortem, should be done only in accordance with domestic law consistent with the international law of human rights.[31]

The need to protect the confidentiality of genetic data concerning an identifiable person has been emphasized not only by the Commission on Human Rights in its resolutions in 2001 and again in 2003 on human rights and bioethics, as well as by the Expert Group on Human Rights and Biotechnology convened by the UN High Commissioner on Human Rights and by the European Group on Ethics in Science and New Technologies, which consider genetic data to be personal and part of the identity of the individual. It is stated that the exercise of an individual's right to health should in no way be limited by her or his genetic characteristics; in the sphere of employment and general insurance including life insurance, genetic characteristics may sometimes legitimately be used to discriminate between individuals in relation to the availability and, if possible, the costs of insurance.[32] Even for

[30] See generally, Vyas, Sreenicasulu, and Kartikeya. 2016. 'Biotechnology, Genetic Databases and Human Rights'

[31] Article 12: 'Collection of Biological Samples for Forensic Medicine or in Civil, Criminal and Other Legal Proceedings' of the Declaration on Human Genetic Data.

[32] UNHCHR's Expert Group on Human Rights and Biotechnology, Health and Human Rights. 2002. 'Expert Group on Human Rights and Biotechnology Convened by the UN High Commissioner for Human Rights:

employment purpose, it is said that genetic information should not be the criteria or the same should not be used in any way for identity. All human beings must be respected, irrespective of their genetic character-istics, according to the mandate of the UN Declaration on Genetic Data.

The Charter of Fundamental Rights of the EU, an instrument adopted when the development of biogenetics had already shown all its adverse potentialities, is of particular relevance in this context. It states that any discrimination based on any ground such as sex, race, color, ethnicity or social origin, genetic features, language, religion or belief, political or any other opinion, membership of a national minority, property, birth, disability, age, or sexual orientation must be prohibited. While the principle of nondiscrimination is well crystallized in the framework of general international law, it is at the basis of human rights protection, being inherent to the very idea of human dignity.[33]

The UN Declaration on Human Cloning was adopted in 2005, under which, member states are required to prohibit all forms of human cloning inasmuch as they are incompatible with human dignity and the protection of human life. The members are expected to adopt measures to prohibit the application of genetic engineering techniques that may be contrary to human dignity.[34]

UNESCO adopted the Universal Declaration on Bioethics and Human Rights in 2005, which is the latest among the international conventions regulating human rights issues involved in biotechnology.[35] The declaration intends to further address human rights and ethical concerns involved in biomedical research. It postulates respect of human dignity and human rights.[36] The declaration recommends independent,

---

Conclusions on Human Reproductive Cloning', The President and Fellows of Harvard College on behalf of Harvard School of Public Health/François-Xavier Bagnoud Center for Health and Human Rights, 6(1): 153–9, available at http://www.jstor.org/stable/4065318 (last accessed 20 June 2016).

[33] Francioni and Scovazzi (eds). 2006. *Biotechnology and International Law*, p. 323.

[34] See Plomer and Torremans. 2009. *Embryonic Stem Cell Patents, European Law and Ethics*, p. 211.

[35] See Hocking, Barbara Ann. 2009. *The Nexus of Law and Biology*, p. 175. Surrey, England: Ashgate Publishing Limited.

[36] Hocking. 2009. *The Nexus of Law and Biology*, p. 182.

multidisciplinary, and pluralist ethics committees to be adopted for over-seeing the research in biotechnology, biology, and medicine. It states that any decision or practice within the scope of the declaration should be founded on the recognition of the primacy of a person, which will prevail over the sole interest of science or society.

The Agreement on Trade-Related Aspects of Intellectual Property Rights (TRIPS) also affirms that inventions that may be detrimental to the dignity and health of human beings, animal, and plants, or to the environment, may also be excluded from the scope of protection. Few researches in biotechnology involving human genetic material, stem cells, and human embryos are controversial and contextual and have an impact on the notions of human dignity. Surgical, therapeutic, and diagnostic methods for the treatment of human beings are prohibited under the agreement from private protection, while leaving them for open use for the public.[37] The agreement believes that certain researches and inventions of biotechnology such as human cloning, which alter the genetic identity of human beings, would go against human dignity. People who are advocates of human rights say that the agreement excludes human cloning from commercial protection as a matter of respect for human rights and its values.

The application of international provisions on patentability of human genetic elements would affect the biotechnological research and its benefits, conflicting with the right to health of any person as per general international law. Any technological development has two sides: its making and its implementation. Similarly, in the biotechnology and genetic research development also, there are different sides to what it appears to us at the outset. It is essential that in the context of genetic biotechnology, while attaining the greatest improvements in healthcare, food supply, and industrial development, it is also important to respect the primacy of human dignity and human rights as envisaged under the international norms on human rights. It is argued that classical human rights instruments are based on a "thick" conception of the subject of human rights and human dignity as an embodied, autonomous person. By contrast,

---

[37] However patenting of surgical, therapeutic, and diagnostic methods for the treatment of plants are considered not to go against the ethical standards of the society.

a thin, religiously inspired conception of the subject of human dignity is also potentially engaged in modern human rights instruments in the fields of genetics.[38]

Protection of Human Rights under the EU Directive on Biotechnology

Human dignity is expressly mentioned as a guiding interpretive principle in the EU directive on biotechnological inventions. Perhaps, the drive for a comprehensive mechanism in the EU on the regulation of biotechnology got accomplished with the adoption of a directive on the legal protection of biotechnological invention.[39] The directive acknowledges the latest advancements in biotechnology, such as human genetic research, embryo cloning, and stem cell research, and, at the same time, is cautious about the human rights issues involved.[40] The directive empowers the European Group on Ethics in Science and New Technologies to evaluate ethical and human rights concerns in biotechnology from time to time.[41] The directive prohibits the following inventions in the field of human genetics for being violative of human dignity:

1. The human body at various stages of its formation and development and the discovery of one of its elements including the sequence or partial sequence of a gene.
2. Inventions that are contrary to public order and morality, such as:
   a. processes for cloning of human embryos;
   b. processes for modifying the germ-line genetic identity of human beings;
   c. use of human embryos for industrial or commercial purposes;
   d. processes for modifying the genetic identity of animals that are likely to cause them suffering without any substantial

---

[38] Plomer and Torremans. 2009. *Embryonic Stem Cell Patents, European Law and Ethics*, p. 204.

[39] The EU Directive on the legal protection of biotechnology inventions, 1998.

[40] Sreenivasulu. 2012. 'Patenting of Biotechnology and Human Rights'.

[41] Article 7 of the directive: 'The Commission's European Group on Ethics in Science and New Technologies evaluates all ethical aspects of biotechnology.'

medical benefit to humans or animals and also the animals resulting from such processes.[42]

A lot of intellectual efforts have gone into the framing of the directive in the form of the number of drafts presented before the European Parliament and a number of committees having worked on the socio-economic, ethical, and the human rights issues in biotechnology. By stating that the human body, including germ cells and notably the human embryo, and all elements of the human body should not be considered patentable inventions, the directive makes it clear that human rights and human values cannot be overweighed by the developments in research, leading to private monopoly including intellectual property such as patents. The directive intends to ensure that human body effectively remains unavailable and inalienable, and that human dignity is thus safeguarded.[43]

However, shortly after the adoption of the directive, in the case of *The Kingdom of the Netherlands v. European Parliament and Council*,[44] a petition was filed before the European Court of Justice (ECJ) challenging the validity of the directive. The petition contended that the directive was contrary to fundamental rights and human dignity. However, the ECJ rejected the contention while stating that grant of patents in biotechnology including patents on isolated elements of human body would undermine human dignity. The court, while reading and confirming the directive, went on to say that human body is effectively unavailable and inalienable and human dignity is very much safeguarded under the directive. It was held that the directive encourages patents in isolated and

---

[42] Article 5 of the Directive states as follows:

1. The human body, at the various stages of its formation and development, and the simple discovery of one of its elements, including the sequence or partial sequence of a gene, cannot constitute patentable inventions.
2. An element isolated from the human body or otherwise produced by means of a technical process, including the sequence or partial sequence of a gene, may constitute a patentable invention, even if the structure of that element is identical to that of a natural element.
3. The industrial application of a sequence or a partial sequence of a gene must be disclosed in the patent application.

[43] Francioni and Scovazzi (eds). *Biotechnology and International Law*, p. 296.
[44] Case 377/98, ECR I-7079, 2001.

purified elements of human body such as genetic material and does not encourage natural elements of human body or such genetic material for patents. The progress of biomedical research through biotechnology and genetic research is essential for the sake of humankind, and as such the directive does exactly that and nothing more. Therefore, the purposes of the directive should not be misunderstood or misinterpreted in any other ways and means. However, interestingly, the court did not debate whether human embryonic research and patents on the same would violate human dignity or not.

The use of information derived from genetic testing for recruitment purpose would, perhaps, go against the right to dignity, equality, and equal opportunity and other such human rights. However, genetic testing is widely used in certain countries of the world, such as the US and the EU including the UK as a means of assessing the ability of individuals to perform certain work. In *Mile M v. Commission of the European Communities*,[45] the ECJ stated that a assessment of fitness to perform given work may be based not only on the existence of actual disorders but also on a medically justified prognosis of future disorders capable of jeopardizing the normal performance of the individual in question in the foreseeable future, thus implicitly accepting the idea that such future disorders may also be assessed by taking recourse to genetic screening. Further, in *A v. Commission of European Communities*,[46] the same position was taken by the ECJ.

The ECJ, in these cases, also maintained that, although a person may not be forced to undergo an AIDS screening test in the context of a prerecruitment medical examination as an employee of a community institution, the institution cannot be obliged to take the risk of recruiting such person who does not consent to undergo testing. In *X v. Commission of the European Communities*,[47] the idea of prerecruitment genetic tests was further supported and considered as legitimate. However, in the Indian context neither such genetic testing for employment or recruitment purpose could be prescribed nor does the current jurisprudence support genetic testing for the said purpose as such as it

[45]   Case 155/78, ECR 1797, 1980.
[46]   Case T-10/93, ECR II-179, 1994.
[47]   Case C-404/32, ECR I-4737, 1994.

would go against the Articles 15 and 16 of the Indian constitution that guarantee the right against discrimination of any sort and right to equal opportunity in employment.

In comparison to the US, human rights concerns in biotechnology are well-founded in the EU. The Convention for the Protection of Human Rights and Fundamental Freedoms, the Convention on Human Rights and Biomedicine, and the Directive on Biotechnology[48] specifically address human rights concerns involved in biotechnological inventions. Although human cloning is not directly prohibited in the EU, certain inventions in the field of human genetics are prohibited, since these are considered to be grossly violative of human rights. The International Declaration on Human Genome and Human Rights has been a major progress in the way of balancing the human rights issues with the research in human genetic material. Human Cloning Prohibition Act, 2003, passed by the Congress of the US has been a measure in addressing the issues of human dignity in the biotechnology and genetic research.

Indian Response to Human Rights Concerns in Biotechnology

India is considered a developing country, with a larger and wider approach toward human rights in general and values associated with life and living beings in particular. The legislature, judiciary, and executives is greatly sensitized toward human rights issues, but there is lack of dialogue and codification regarding biotechnology research and connected human rights. Although there has not been much literature and judicial dictum on the human rights–related issues of biotechnology, there have been certain initiatives by the government.[49] The Indian Council of Medical Research (ICMR), in its guidelines, addresses the issues of human rights and dignity while using human beings for biomedical research, including genetic research. Furthermore, stem cell research guidelines were formulated under the aegis of the Department

---

[48] See Plomer and Torremans. 2009. *Embryonic Stem Cell Patents, European Law and Ethics*, p. 19.

[49] Sreenivasulu. 2008. *Biotechnology and Patent Law*, p. 207.

of Biotechnology for regulating stem cell research and related human rights issues.

Human Genetic Research and Human Rights

Assumptions and convictions concerning human rights have been at the center of debates on biotechnology, which play a pivotal role in the formulation of law and policy of biotechnology. Under the international law, the deep seabed, Antarctica, and the moon and other celestial bodies are considered to be the common heritage of mankind. The UDHGHR intends to confer greater solemnity and the status of common heritage of mankind to human genome to avoid material exploitation, as it is physically, tangibly appreciable.[50] Furthermore, it is stated under the convention that benefits from advances in biology, genetics, and medicine, concerning the human genome, will be made available to all with due regard for the dignity and human rights of each individual.[51] Research in genetics progressed to the extent of producing human organs to transplant into the body of needy. Cloning of human genetic material such as cells, genes, and DNA to produce human organs for transplantation has become a reality, irrespective of the ethical considerations involved in cloning a part of the body. Biotechnology is poised to serve the purpose of the needy. Research in human genetics is often said to tinker with human dignity while intending to unravel the genetic structure of human body in order to cure genetic diseases. These research initiatives have forced the human rights to adjust to the dynamism in the genetic research.[52] Human genomic research involves manipulating the genetic setup of the individual, which is being criticized by several human rights forums and associations.

Application of human genetic research has reached the heights and breached unimagined barriers. For instance, embryo research involves

[50] Human genome is defined as the heritage of humanity under the UDHGHR.

[51] Francioni and Scovazzi (eds). 2006. *Biotechnology and International Law*, p. 303.

[52] Sreenivasulu, N.S. 2012. 'Biotechnology, Patents and Human Rights', *International Journal of Law and Policy Review, School of Technology Laws*, I (2): 146–58.

manipulation of such embryos which are capable of developing into a complete human being. Manipulating the embryo or fetus and removal of stem cells of such embryos in the initial stages of the development of a human being are strongly objected as they violate the right to life of the embryo, which might become a human being if it is left to grow. Genetic research further brought into existence the probable possibility of begetting children through assisted reproduction to the benefit of infertile couples.[53] The Human Embryology Authority in the UK regulates human embryo research in the light of latest advancements in this field, which involves serious ethical and human rights concerns. In view of the fact that human embryo research involves serious ethical and human rights concerns, the authority assesses the human right issues in cloning and using human embryos for therapeutic purposes and assesses the consequences.

Patenting Human Genetic Material and Human Rights

It is believed that biotechnology is capable of manipulating any living being, including human beings. In particular, biotechnology is known for marshaling the genetic material of any living being. It is capable of isolating, removing, suppressing, or incorporating genes from one human body into another.[54] Furthermore, it has provided the facility and technique to identify, isolate, and commercially produce proteins, hormones, and enzymes that a human body produces naturally. It was felt that the alteration of the natural setup of human body would go against basic values of human existence and human rights. It is pointed out that patenting of human genetic material such as cells, DNA, and genes amounts to owning human body or human life as private property. Owning human beings or human life would amount to slavery,

[53] It is even possible to have children with desired characteristics through assistance of biotechnology. In the light of increasing number of fertility clinics and also increased research in the field of embryology, the Government of United Kingdom enacted Human Fertilization and Embryology Act, 2008. The act establishes Human Fertilization and Embryology Authority to monitor the working of fertility clinics in helping parents to have children and regulate assisted reproduction.

[54] See Sreenivasulu, N.S. 2005. 'Patenting Genetically Modified Life Forms: Legal Issues and Challenges', *Indian Bar Review*, 32(3&4): 485–98.

which is against the human dignity and grossly violate human rights. The moral basis and meaning of human dignity are variable and indeterminate and the attribution of human dignity as a ground for human rights reflects a diversity of philosophical, moral, and religious traditions and cultures.[55] Debates on violation of human rights and human dignity by manipulating human body and in obtaining patents on human body parts started quite early. It got intensified with the patenting of living beings. Violation of human values and rights takes place at two different levels here. One type of violation takes place in the process of manipulation of living beings and human bodies and the second in holding patents on the manipulated bodies. Manipulation of human body by way of isolation or removal of genetic material such as genes and DNA or incorporation of foreign genetic material into the body is possible through genetic engineering. Eventually patents that are being claimed on human genetic material give rise to heated debates on the ethics involved in manipulating human body and morality in owning human genetic material.

A major advancement in biotechnology took place in the 1990s. In the US, there has been a major upsurge in the research on human genetic material, resulting in a number of patents on human DNA and genes. In *Amgen v. Chugai*,[56] the claim was for a genetic material, in particular DNA. Although the patent office did not have many objections as it had anticipated patent applications on human genetic material having seen the progress of the biotechnology industry. However, there was an apprehension in the circles of patent office and the patent advocates that patenting genetic material may be opposed by the advocates of human rights and nongovernmental organizations (NGOs). The apprehensions became a reality, since the canopy of patent law started encroaching upon the realm of human beings, in particular the sphere of human rights tinkering with the human dignity. In fact, human dignity is the core concern while doing biotechnological[57] R&D. As the fundamental

[55] Plomer and Torremans. 2009. *Embryonic Stem Cell Patents, European Law and Ethics*, p. 204.

[56] 927 F.2d 1200, 18 USPQ 2d 1016 (Fed. Cir. 1991) (henceforth 'Amgen').

[57] See Francioni and Scovazzi (eds). 2006. *Biotechnology and International Law*, p. 373.

opponents of patents on genetic material felt that granting of patents on human genetic material would violate the dignity of the human beings as such, the research in human genetics would tinker with the integrity of the human being.[58]

Going by the strict language of the various human rights conventions that guarantee human dignity, integrity, and self-sustenance, human genetic research would definitely disturb the notions and standards of human rights. However, as the patent office was poised to extend the scope of the patent law toward the growth of human genetic research, having recognized the potential of such research particularly in the field of medicine, the perceptions of human rights in the stricter sense were not being emphasized. Rather, the approach of the patent office in granting patents on human genetic material in *Amgen* case promoted the advocates of patents on genetics to alter the fabric of human rights notions in order to see that there is no negative impact of human rights standards on the patenting of human genetic material. Having set the platform ready with its decision in *Amgen* case, the US Patent Office entertained many such patent applications on human genetic material where little or no emphasis was given to the human rights considerations. In *In re Bell*,[59] DNA was claimed whereas in *In re Deuel*,[60] the invention related to an isolated and purified DNA was claimed. In *In re O'Farrell*,[61] the invention was a method to produce a foreign protein in a transformed species of bacteria.

Since the beginning of the 1990s, a large number of genes and DNA has been patented. Every day several genes, DNA, and other human genetic material are being patented. Thousands of patent applications are pending before the US Patent Office claiming different biotechnology inventions, a majority being claims on human genetic material.[62] Reportedly, in 2001 alone, the US Patent Office awarded 20,000 gene patents and another 25,000 were pending.[63] Most of the genes and DNA patented are coding some proteins that are useful in the medical

---

[58] Sreenivasulu. 2012. 'Biotechnology, Patents and Human Rights'.
[59] 991 F.2d 781 (Fed. Cir. 1993).
[60] 51 F.3d 1552, 1559, (Fed. Cir. 1995).
[61] 853 F.2d 894 (Fed. Cir. 1988).
[62] Sreenivasulu. 2008. *Biotechnology and Patent Law*, p. 205.
[63] Albright, M. 2002. 'The End of the Revolution', *Gene Watch*, 15.3.

field in the preparation of some kind of medicine, drug, or vaccine. In the EU, for the first time in the *Relaxin* case,[64] the European Patent Office (EPO) happened to dwell on the human rights and ethics involved in the patenting of human genetic material. The claim was for a gene coding for a hormone called relaxin which was produced in the body of women during the time of delivery to relax the body. The patent was opposed on various grounds[65] including that patenting of human gene offends human dignity, as it involves extracting tissue from the human body. It was opined that patenting of human genes amounts to modern-day form of slavery and in the present case it involved using tissues of human body for commercial enterprises. All conventions at the international level and the domestic laws of all countries consider slavery as brutal and barbaric practice which is against the notions of a civilized society.

However, the case argued that extracting tissue from the human body is a standard practice in medical procedure and the same may not offend human dignity. Responding to the argument that patenting gene amounts to slavery of human beings, it was held that since patenting of human genes does not grant any rights on individual human beings, there is no question of slavery. The court concluded that patenting of genes does not amount to patenting of human life; hence, it does not grant any right over human beings to the entirety. Nevertheless, the case has left far-reaching impact on the debate on human rights issues in genetic research and patenting of the same. Furthermore, in *Novartis*,[66] it was held that genetic material, such as cells and parts thereof, are microorganisms and therefore, are patentable,[67] much to the objections of the advocates of human rights. In both the decisions, it seems that the potential use of the biotechnology inventions had taken precedence over the human rights concerns.

[64] (1995) Official Journal of the European Patent Office 388; (1995) E.P.O.R. 541.

[65] Swaminathan, K.V. 2000. *An Introduction to the Guiding Principles of Patent Law*, p. 356–7. New Delhi: Bahri Brother.

[66] *Novartis AG v. IVAX Pharmaceuticals UK Ltd* [2007] All ER (D) 252. EPO Technical Board of Appeals decision 20 December 1999.

[67] Sreenivasulu N.S. 2011. *Intellectual Property Rights*, 2nd edn, p. 52. New Delhi: Regal Publications.

In the light of the aforementioned decisions in *Relaxin* and *Novartis* in the EU, addressing human rights issues in biotechnological research was difficult. In the recent past in Europe, there was a major growth in human genetic research and as a result, more and more human genetic material was claimed for patents. It was feared that the trend might lead to patenting of human beings that would grossly violate the human dignity. Furthermore, in *Kingdom of the Netherlands v. European Parliament and Council of the European Union*,[68] the issue of human dignity in the biotechnological research was addressed. The Netherlands, Italy, and Norway argued that appropriation of isolated parts of the human body would reduce living human matter to a means to an end, undermining human dignity; absence of verification and consent of the donor; or the recipient of products obtained by biotechnological means undermines the right to self-determination. The court viewed that natural elements when combined with technical process in an isolated form enabled for industrial application would be eligible for protection. Furthermore, it was added that elements of human body in their natural form are not allowed to be appropriated but the nonnatural and improved forms created through biotechnological means are. The court clarified that there will be no apprehension regarding the dignity of human beings in such cases. Human body effectively remains unavailable and inalienable and human dignity is thus safely guarded.

In the *Omega* case,[69] the ECJ held that fundamental rights of humans form an integral part of the general principles of law, as assured by the court. For this purpose, the court drew inspiration from constitutional traditions common to the member states and from the guidelines supplied by international treaties for the protection of human rights, which the member states have collaborated on or which they are signatories to.[70] The court further stated that the European Convention on Human Rights and Fundamental Freedoms has special significance in this respect.

[68]   Case C-377/98, 2001, ECR-I-07079.
[69]   *Omega Spielhallen v. Oberburgermeisterinder Bundesstadt Bonn*, 2004, Case 36/02, ECR-I-09609.
[70]   Plomer and Torremans. 2009. *Embryonic Stem Cell Patents, European Law and Ethics*, p. 220.

Human Rights Concerns in Research on Human Cell Lines

For the first time, in *John Moore* case,[71] a patent was claimed on a cell line of a human being, igniting the debate on the human rights concerns in genetic research, in particular on patenting human cell lines.[72] John Moore, a leukemia patient from California, underwent a medical treatment for hairy cell leukemia. While treating him, his physicians found his cell lines useful in preparing a specific medicine and obtained patent on the same in 1984. Moore challenged patenting of his cell line in the Supreme Court of California by arguing that patented cells are from his body, which is his property on which only he has rights. Patenting of cell lines of Moore and his contention of property rights on his own body raised serious human rights concerns. The morality of patenting a cell line, which is a part of a human body, was being questioned. It was argued that owning a part of the human body through patent is a gross violation of human dignity and moral standards of the society.

The court rejected Moore's argument and held that there cannot be any property rights over one's body, which is against dignity of human beings. It viewed that patent was not claimed on the natural cell lines in the body but on the isolated cell lines outside the body, which deserved patent. Eventually the court gave green signal to patenting of human cells, genes, and DNA, much to the agitation of the human rights advocates. However, the debate continued and further intensified after the decision of the court.

Again, in 1993, an application was submitted for a patent on human cell lines before the patent office.[73] The cell line was developed from blood, which was found to be useful in research on AIDS and cancer. This time it was not an individual claiming the patent but it was the US government which claimed cell lines of women from the Guaymí tribe of Panama, South America.[74] The US government contended that the patent could be granted, as the cell line is useful in research

[71] *John Moore v. Regents of University of California* 1990, 51 Cal, 3d 120: 271, Cal. Rpte, 146. Decided by the Supreme Court of California, 9 July 1990.

[72] See Sreenivasulu. 2008. *Biotechnology and Patent Law*.

[73] Patent application No. 9208784 A 1, 1993.

[74] Jain, Rajiv and Rakesh Biswas. 1999. *Law of Patents, Procedure and Practice*, 2nd edn, p. 1.36. New Delhi: Vidhi Publishing Co. (P) Ltd.

in AIDS and cancer, two of the most life-threatening diseases in the world. There were protests before the US Patent and Trademarks Office against patenting of cell lines of human beings. NGOs such as Rural Advancement Foundation International (RAFI) and tribal communities strongly objected[75] and questioned patenting a cell of tribal women much against the dignity of the human beings.[76] As the US Government itself was involved in the patent, the opponents criticized the stand of the government in claiming a cell line that would lead to commodification of life. The application claiming the human cell line of tribal women attracted international criticism. Yielding to the high pressure, vehement oppositions, and international criticism, the government was forced to withdraw the patent application on human cell lines. These two cases reveal how human rights are at stake in the patenting of human cell lines.

Human Right Issues in Cloning

Research in biotechnology and genetics took an interesting turn with the discovery of techniques of cloning.[77] Cloning is a general term that explains any procedure that produces a precise genetic replica of a biological object including a DNA sequence, a cell, or an organism. Furthermore, there is another question of fundamental importance related to cloning, that is, the viability of cloning in a human rights discipline. In relation to human rights, the possibility of human cloning represents a violation of the two fundamental principles on which all human rights are based: the principle of equality among human beings and the principle of non-discrimination.

According to Juan de Dios Vial Correa of the Pontifical Catholic University of Chile, cloning of human beings results in upsetting the

[75] RAFI Press Release, 'Patenting the Primitive Anthropological and Ethical Reflections on Indigenous Peoples and the Human Genome Project', 26 October 1993, cited in Foster, Meika. 1999. 'The Human Genome Diversity Project and the Patenting of Life: Indigenous People Cry Out', *Canterbury Law Review*, available at http://www.nzlii.org/nz/journals/CanterLawRw/1999/7. html (last accessed 20 June 2016).

[76] Blenty, Lionel and Spyros M. Maniatis. 1998. *Intellectual Property and Ethics*. London: Sweet and Maxwell.

[77] See http://cnx.org/contents/peFfu7QM@1/Biotechnology---Cloning-and-Ge (last accessed 1 March 2016).

basic principle of parity and equality in all human beings.[78] He thus rejects cloning, arguing that the aspect of selective eugenics or the ability to program character traits influenced by specific genomic sequences "denies the dignity of the person subjected to the cloning and the dignity of human procreation."[79] The prospect of creating successful human clones has thrown open a whole gamut of possibilities; people now ask themselves whether the process will ultimately lead to a "commercialization" of humans, as the power to give life may be "marketed" and people would be turned into "disposable and recyclable products."[80] The questions posed put forward a strong argument for the preservation of human dignity and life created through natural procreation, and though steeped in theology, these arguments cannot be wholly discarded on the basis of scientific justifications.

International Law on Human Cloning

The UDHGHR is the principal text of the UN in the area of human biotechnology and human rights. Together with the guidelines for its implementation adopted in 1999, the UDHGHR sets out a framework for dealing with new human rights issues posed by advances in technology related to the human genome. Article 11 of the UDHGHR prohibits reproductive cloning by terming it as the major example of "practices contrary to human dignity that cannot be permitted."[81] In fact, the debate on human reproductive cloning has influenced the framing of policy responses.[82] The UDHGHR sets out a framework for dealing with new

[78] Sreenivasulu, N.S., Rohan Benarjee, and Arpan Narayan Choudary. 2016. 'Cloning Technology, Public Policy and Human Rights', in N.S. Sreenivasulu (ed.), *Human Rights and Development*, 1st edn, pp. 201–28. Indiana, USA: Penguin-Partridge, Bloomington.

[79] See http://www.vatican.va/roman_curia/pontifical_academies/acdlife/documents/rc_pa_acdlife_doc_30091997_clon_en.html#ETHICAL PROBLEMS CONNECTED WITH HUMAN CLONING (last accessed 1 March 2016).

[80] See http://mitworld.mit.edu/video/239/ (last accessed 1 March 2016).

[81] See Francioni and Scovazzi (eds). 2006. *Biotechnology and International Law*, p. 311.

[82] Plomer and Torremans. 2009. *Embryonic Stem Cell Patents, European Law and Ethics*, p. 38.

human rights issues posed by advances in technology related to the human genome and, in some ways, gives vent to the ethical approach. The Expert Group on Human Rights and Biotechnology, convened by the UN High Commissioner on Human Rights, submitted a report in 2002, emphasizing that cloning for reproductive purposes is an area of biotechnology with the greatest potential for controversy at the moment. An attempt was made to draw a line between reproductive cloning and therapeutic cloning, and more importantly for recognizing "the importance of therapeutic cloning as providing possibilities for preventing and fighting diseases."[83]

The draft of WHO Guidelines on Bioethics also states that cloning for the replication of human individuals as well as deliberate reproductive cloning by embryo splitting are contrary to human dignity and integrity. A similar approach has been adopted by the Ad Hoc Committee on an International Convention against the Reproductive Cloning of Human Beings, appointed by the UN General Assembly in 2001. Further, the draft WHO Guidelines on Bioethics reject reproductive cloning as "morally unacceptable" and "contrary to human dignity" but also state that "non-reproductive cloning research, with the clinical objective of repairing damaged tissue, has important potential benefits and should be encouraged." In spite of such divergent views, the UN General Assembly passed a resolution in 2005, the United Nations Declaration on Human Cloning, which seeks to prohibit all forms of cloning.[84] The declaration calls upon the member states to prohibit all forms of human cloning citing them as against human dignity and the protection of human life. Member states are further called upon to adopt the measures necessary to prohibit the application of genetic engineering techniques that may be contrary to human dignity.[85]

Human Cloning and the EU, and the UK

The stand of the EU as a whole on the issue of human cloning is actually uniform and restrictive. Reproductive cloning is explicitly prohibited

[83] See Sreenivasulu. 2008. *Biotechnology and Patent Law*, p. 207.

[84] The Declaration called on the member states to adopt all measures necessary to prohibit all forms of human cloning inasmuch as they are incompatible with human dignity and the protection of human life.

[85] Francioni and Scovazzi (eds). *Biotechnology and International Law*, p. 335–6.

by Article 3 of the Charter of Fundamental Rights of the EU.[86] The Additional Protocol on "Prohibition of Cloning of Human Beings" to the EU convention for the protection of human rights and dignity of human beings with regard to the application of biology and medicine bans reproductive cloning as it constitutes an act "contrary to human dignity."[87] The European Parliament, in its Resolution on Human Cloning of 7 September 2000, specifically stressed on the fact that it did not perceive "any difference between cloning for therapeutic purposes and cloning for the purposes of reproduction."[88] This somewhat stricter view of the European Parliament is not shared by the European Group on Ethics in Science and New Technologies, which seeks to create a clear distinction between "cloning aimed at the birth of identical individuals and non-reproductive cloning limited to the in vitro phase."[89] While the group univocally condemns reproductive cloning, it concedes the fact that "the philosophical and scientific debate on cloning is open and it should be stressed that prohibition of a scientific technique may prevent important discussions about human genetics."[90]

The UK prohibits reproductive cloning; it has specific legislations that allow therapeutic cloning to be carried out for the purposes

[86] Charter of Fundamental Rights of the European Union, (2000/C 364/01) Official Journal of the European Communities, available at www.europarl.europa.eu/charter/pdf/text_en.pdf (last accessed 1 March 2016).

[87] Plomer and Torremans. 2009. *Embryonic Stem Cell Patents, European Law and Ethics*, p. 38.

[88] The ideology of the European Parliament is further strengthened by the fact that it calls for "adoption of human artificial insemination techniques that do not result in excess number of embryos," since it is opposed to conducting scientific research even on superfluous embryos.

[89] This view of the group was apparent in its report, *Citizen's Rights and New Technologies: A European Challenge* on the "Ethical Aspects of Research Involving the Use of Human Embryo," a programme of the EU. See Busby, Helen, Hervey, Tamara, and Mohr, Alison. 2008. "Ethical EU law? The influence of the European Group on Ethics in Science and New Technologies," *European Law Review*, 33: 803–42.

[90] Sreenivasulu, N.S., R. Benarjee, and A.N. Choudary. 2015. 'Human Cloning: Legal and Policy Concerns', in N.S. Sreenivasulu (ed.), *Human Rights and Development*, 1st edn, pp. 155–75. Bloomington, Indiana, USA: Penguin-Partridge Publications.

of medical research and treatment. The Human Fertilization and Embryology Act of 1990 is a landmark legislation in the field of cloning. This act, based on the "Warnock Report," confers a proportional moral status to the embryo, and provides for a system of licensing by the Human Fertilization and Embryology Authority for any research on cloned embryos. Keeping human dignity and such other values that the human rights endorse, embryo research is permitted only up to 14 days after fertilization of the embryo or the appearance of the "primitive streak," which marks the development of the embryo into a moral status comparable to a living individual.

In a legal challenge initiated by Pro-Life Alliance in 2001[91] against the act, in R (Bruno Quintavalle) v. Secy. of State for Health it was viewed that human embryos, whether natural or artificially created, would have equal status under the human rights law as probable and prospective human beings. More importantly, this case led to proactive steps being taken by the UK Legislature to plug the loopholes in the 1990 Act by introducing the Human Reproductive Cloning Act of 2001 and the Human Fertilization and Embryology (Research Purpose) Regulations, 2001, that regulate and prohibit human cloning and allow only therapeutic cloning for scientific research and medicinal purposes.

Policy Perspectives in the US

As 13th amendment to the US constitution prohibits slavery in human beings, the same was taken to restrict human cloning and patenting of human beings. Besides, the US judiciary believes that cloning of human beings and patenting the same are grossly violative of human rights and are totally excluded under the US constitution as well as under the patent law. Both the judiciary and the patent office state that cloning of human being violates the inherent dignity of human life and patenting of human being is equal to slavery.

This stand was highlighted in the Pioneer Hibred International case[92] where it was held that human cloning is prohibited as it is against public

[91] [2003] UKHL 13. Also see R (Josephine Quintavalle) v. HFEA [2005] UKHL 28.
[92] Pioneer Hibred International v. Holden Foundation Seeds Inc., 35F, 3d. 1226.3, USPQ, 2d. 1385 (8th Cir. 1994).

policy and human dignity. However, the policy in this regard raised certain doubts, since processes or methods for human cloning were held patentable. At this juncture, it was felt that the human rights concerns vehemently demanded for the total ban of human cloning. The US brought a legislation in 2003[93] on the prohibition of human cloning which prescribed 10 years' imprisonment for executing, attempting, facilitating, aiding, and assisting in cloning human beings. However, for therapeutic reasons, cloning of human beings was allowed with a strict ban on the cloning for reproduction purpose. Unlike the UK, where cloning for research purposes is allowed by law, in the US cloning for both "reproductive" and "nonreproductive purposes" is prohibited. A sole exception is made with regard to allowing stem cell research on "existing stem lines."

Status in India

India, like most other nations, is opposed to reproductive cloning asserting that it is unacceptable and violative of human values and dignity, but supports research on stem cells saying that the new technology could be used to fight certain diseases. Stem cell research in India is regulated from the human rights and ethical perspectives by the guidelines issued by the ICMR and the DNA Safety Guidelines brought out by the Government of India. Although the guidelines allow research on embryonic stem cells, no embryo can be created for the sole purpose of obtaining stem cells.[94] The National Apex Committee for Stem Cell Research and Therapy (NAC-SCRT) and an Institutional Committee for Stem Cell Research and Therapy (IC-SCRT), registered under NAC-SCRT,[95] would analyze the scientific, technical, ethical, legal, and human rights issues in embryonic stem cell research. The primary reason why cloning is prohibited despite allowing stem cell research may be attributed to the philosophy that cloning could be misused to develop complete human

---

[93] Human Cloning Prohibition Act, 2003.

[94] See http://cellnews-blog.blogspot.com/2007/11/draft-guidelines-for-stem-cell-research.html (last accessed 1 March 2016).

[95] Sreenivasulu, Benarjee, and Choudary. 2015. 'Human Cloning: Legal and Policy Concerns'.

beings and then used against the humanity. Therefore, with due respect to human rights philosophy, research is allowed only on the "superfluous embryos" and cloning of embryos for research purposes is not allowed.

Biotechnology and Food Security: GM Foods to Ensure Right to Food

In the field of agriculture, biotechnology plays an important role in yielding greater amounts of food crops for mass consumption and cash crops for industrial use; yielding improved varieties of existing crops with reduced vulnerabilities; and more.[96] As population grows, the demand for agricultural production grows and also the need to ensure right to food to everybody. In earlier times, agriculturists used different methods to increase crop production, such as tissue culture, crop budding, and so forth. In the 1960s, the Green Revolution played an important role in achieving self-sufficiency in foodgrain production.[97] The relevance of biotechnology arises from the need for the infusion of a new round of technological change in the Indian agriculture. According to reports in 2012, India is home to the largest number of malnourished children, where approximately 3,000 children die daily due to hunger and malnutrition.[98] With rising population, such problems will get

[96] Sreenivasulu. 2008. *Biotechnology and Patent Law*, p. 6.

[97] Rao, N. Chandrasekhara and S. Mahendra Dev. 2010. *Biotechnology in Indian Agriculture: Potential, Performance and Key Concerns*, p. 28. New Delhi: Academic Foundation.

[98] Reuters. 2012. 'India's Hunger Shame: 3,000 Children Die Every Day, despite Economic Growth,' 16 February, available at http://worldnews.nbcnews.com/_news/2012/02/16/10424930-indias-hunger-shame-3000-children-die-every-day-despite-economic-growth?lite (last accessed 2 March 2016). See also Press Information Bureau, Government of India, Prime Minister's Office. 10 January 2012. 'HUNGaMA: Fighting Hunger and Malnutrition,' the HUNGaMA Survey Report, Hyderabad: Naandi Foundation, available at http://pib.nic.in/newsite/PrintRelease.aspx?relid=79457 (last accessed 21 June 2016); See also Singh, Manmohan. 2012. *PM's speech at the release of HUNGaMA (Hunger and Malnutrition) Report*, 10 January, available at http://pmindia.nic.in/speech-details.php?nodeid=1125 (last accessed 19 November 2012).

aggravated. In order to achieve the millennium development goals and fulfill domestic obligations (constitutional and statutory) of food and nutritional security and economic growth, biotechnology is of considerable importance to the society at large.

The Department of Biotechnology under Ministry of Science and Technology, Government of India, in its vision statement on "Biotechnology in India," states that it will strive to achieve nutritionally enhanced crops with high yield with the help of agricultural biotechnology, which is a precursor for ensuring food security and the right to food. Food security exists when all people, at all times, have physical, social, and economic access to sufficient, safe, and nutritious food that meets their dietary needs and food preferences for an active and healthy life. Essentially, food security can be described as a phenomenon related to individuals. It is the nutritional status of the individual household member that is the ultimate focus, and the risk of that adequate status not being achieved or becoming undermined. Food security has been a major developmental objective in India since the beginning of planning period in 1951. India achieved self-sufficiency in foodgrains in the 1970s and has sustained it since then. However, the achievement of foodgrain security at the national level did not percolate down to households and the level of chronic food insecurity is still high. Over 225 million Indians remain chronically undernourished. In 2000–1, about half of the rural children below five years of age suffered from malnutrition and 40 percent of adults suffered from chronic energy deficiency.

In 2000–1, it was found that about half of the rural children below 5 years of age were malnourished and 40 percent of adults suffered from chronic energy deficiency. Such a high level of wastage of human resources should be a cause for concern. However, rapid strides were made in agriculture and allied sectors after industrialization, in particular in the post-independent India. From being a net importer of foodgrains, not only has the country achieved food security through domestic production, but it has also become a regular exporter of several commodities. Statistically speaking, India has gone up from producing foodgrain of about 50 million tons in 1950 to 241 million tons in 2010–11. In spite of such spectacular achievements, the road to ensuring and maintaining food security in the years to come is challenging with the deceleration in the availability of foodgrains.[99] Production and availability of foodgrains is increasingly required for ensuring food security, which is a fundamental right. In the era of biotechnology, genetically modified (GM) foods are recommended for ensuring food security for the masses. India being a country with two-thirds of the population depending upon agriculture, the use of biotechnology would not only provide better results for the agriculturalists but also benefit the consumers and the public at large.[100]

Biotechnology, GM Crops, and Food Security

The concept of food security cannot be easily defined. At various points over the past 30 years, food security has taken on different meanings depending upon the location of the policy debate. In the mid-1970s, policymakers at the global level were emphasizing on the availability of food and its supply. A decade later, in the 1980s, the emphasis had shifted to access to food and the importance of well-being that food security brings with it. In the 1990s, food security formed part of the overall focus of policymakers on the alleviation of poverty.[101] The definition of food security that is most commonly used is that from the Rome Declaration,[102] which considers the fundamental right to adequate food and defined food security as the right of everyone to have access to safe and nutritious food, consistent with the right to adequate food and the fundamental right of everyone to be free from hunger, or put more simply, access to food for a healthy life by all people at all times. The Food and Agriculture Organization (FAO) International Treaty on

[99] Parliamentary 37th Standing Committee on Agriculture. 2012. 'Cultivation of Genetically Modified Food Crops: Prospects and Effects,' 37th Report of the Parliamentary Standing Committee on Agriculture, presented to the 15th Parliamentary session on 9 August 2012, available at http://164.100.47.134/lsscommittee/Agriculture/GM_Report.pdf (last accessed 21 June 2016).

[100] In terms of economy as well, it constitutes almost 20 percent of our GDP and employs almost 60 percent of India. With every five-year plan, concerted efforts are made by the government to spur the growth of the sector.

[101] Francioni and Scovazzi (eds). 2006. *Biotechnology and International Law*, p. 261.

[102] World Food Summit was held at Rome in 1996.

Plant Genetic Resources for Food and Agriculture (ITPGRFA), also known as the Seed Treaty, along with the Convention on Biological Diversity talk about sustainable development of biological and plant genetic resources toward ensuring food security. The FAO has also continued and extended its close collaboration with the UN Special Rapporteur on the Right to Food on matters of agro-biotechnology, food security, and ethics in food and agriculture. It is interesting to note that the International Assessment of Agricultural Science and Technology for Development (IAASTD) report states that small-holders with access to land, markets, and sufficient resources are the best way to ensure food and livelihood security. The report sees no significant role for GM crops in ensuring food or livelihood security for farmers.[103] Furthermore, the UN Special Rapporteur on the Right to Food, Oliver D'Schutter, in his report to the UN in March 2011, has stated that agro-ecological approaches with low external inputs, which empower farmers and build their knowledge and skills, can effectively increase productivity at the field level, reduce rural poverty, improve nutrition, and help adapting to climatic changes.

Experience with Various Commercially Cultivated GM Food Crops in the US

In the US, soybean and corn are the two major GM crops, often assumed to be food crops, but actually going into nonfood industrial use, livestock feed, and for fuelling automobiles. In 2001, biotechno-logical varieties accounted for about 26 percent of corn, 68 percent of soybeans, and 69 percent of cotton plantation.[104] Today, more than 90 percent of corn, canola, soybean, and beet sugar crops in the US

[103] Mclntyre, Beverly D., Herren, Hans R., Wakhungu, Judi, and Watson, Robert T. 2009. 'Agriculture at Crossroads: International Assessment of Agricultural Knowledge, Science and Technology for Development,' Science and Technology for Development IAASTD, US, available at http://www.unep.org/dewa/agassessment/reports/IAASTD/EN/Agriculture%20at%20a%20Crossroads_Global%20Report%20(English).pdf (last accessed 2 March 2016).

[104] Chaturvedi, Sachin and S.R. Rao. 2004. *Biotechnology and Development: Challenges and Opportunities for Asia*, p. 295. New Delhi: Academic Foundation.

are genetically modified.[105] Herbicide-tolerant soybean (GM HT soybean) and corn have not increased yields any more than conventional methods that rely on commonly available herbicides. Furthermore, insect-resistant Bt corn varieties have provided an average yield advantage of just 3–4 percent compared to typical conventional practices, including synthetic insecticide use. Per acre corn production in the US has increased by 28 percent since the early 1990s. Genetic Engineering (GE) is responsible for only 14 percent of that increase, that is, only 4 percent of the total US yield increase. Meanwhile, non-GE plant breeding and farming methods have increased yields of major grain crops by values ranging from 13 to 25 percent. Bt corn varieties, engineered to protect plants from either the European corn borer or corn root worm, averaged over 13 years since 1996 when it was first commercialized, resulted in around 0.2–0.3 percent operational yield increase per year.[106] Studies by United States Department of Agriculture (USDA) scientists have shown that the yield benefits of insect-resistant crops depend obviously on pest infestation in a given season.[107] On the other hand, in trials of HT soybean, "yield drag" effects were noticed, which were adversely impacting yields.[108] Examining the food security situation of the US and Brazil, which have adopted GM crops on a massive scale, it is evident that the situation has worsened after the introduction of GM crops, though in Argentina, it has remained the same. Clearly, these crops are not meant to address food security or

---

[105] See http://www.gmwatch.org/index.php/news/archive/2014/15574-the-gmo-fight-ripples-down-the-food-chain (last accessed 10 August 2014).

[106] Gurian-Sherman, Doug. 2009. *Failure to Yield: Evaluating the Performance of Genetically Engineered Crops*, report, April. Cambridge, MA: Union of Concerned Scientists.

[107] Fernandez-Cornejo, Jorge and McBride, William D. 2002. *Adoption of Bioengineered Crops*. Agricultural Economic Report No. 810, ERS USDA, 2002, US Department of Agriculture, Economic and Research Service, Washington, USA.

[108] Benbrook, C. 1999. 'Evidence of the Magnitude and Consequences of the Roundup Ready Soybean Yield Drag from University-Based Varietal Trials in 1998,' Ag BioTech InfoNet Technical Paper Number 1, 13 July. Sandpoint, Idaho: Benbrook Consulting Services, available at http://www.mindfully.org/GE/RRS-Yield-Drag.htm (last accessed 21 June 2016).

hunger but to fill the coffers of agri-business corporations whose profits during the same period increased. In the US, in 2011, according to US Economic Research Service, 17.9 million households were food insecure (constituting 14.9 percent of the total American households that were food insecure) at some point in the year.[109] This means that an unprecedented 50.1 million people (1 in every 6 Americans) live in food-insecure households in a nation that has the largest area under GM crop cultivation in the world, after commercializing crops with this controversial technology way back in 1996. Food insecurity has increased to 15 percent of the population from 12 percent during 1995, and since then there has been a consistent increase. Despite adoption of GM technology at the massive scale, the US seems to unable to stem the increasing hunger in the country. It is estimated that USD 322 billion has been paid in farm subsidies during the period of 1995–2014.[110] The biotechnology and GM crops are not even completely acceptable in the US, with doubts and concerns being raised by different stakeholders, including consumers.[111] In Brazil, it is seen that improvements in food security indicators have actually decelerated in the period of expansion of GM crops, compared to the earlier years. Experience shows that in developing countries that have adopted GM crops that there is no direct correlation between improvements in food security and GM crops.

European Experience with GM Foods

The EU follows a stricter and formal regulatory structure for the regulation of genetically modified organisms (GMOs) and GM food, while, at the same time, ensuring not only the right to food but healthy food with food security to its citizens. Quite a few regulations have been brought into force in Europe including specific regulations on

[109] 'Food Security Status of US Households in 2011,' USDA Economic Research Service, available at http://www.ers.usda.gov/topics/food-nutrition-assistance/food-security-in-the-us/key-statistics-graphics.aspx (last accessed 10 September 2014).

[110] See https://farm.ewg.org/ (last accessed 21 June 2016).

[111] See http://www.gmwatch.org/index.php/news/archive/2014/15574-the-gmo-fight-ripples-down-the-food-chain (last accessed 10 August 2014).

GM food,[112] including a regulation concerning tracing and labeling of GMOs,[113] and the directive on the deliberate release of GMOs into the environment.[114] A regulation was adopted requiring the mandatory labeling of all foodstuffs containing GM ingredients above a threshold of 1 percent. In 1997, new regulations for food were introduced in Europe, which introduced pre-market approval system for novel foods including GM foods,[115] in which a manufacturer or importer must show that commercialization of the GM food does not pose a risk to human health or to the environment. In Europe, unlike the US, the process of production of food becomes important. Separate rules exist for GM food, to ensure rights of the consumers to choose between regular food and GM food. Countries such as Japan, Australia, and New Zealand follow the European path by adopting similar policy for the regulation of GM food. It is perceived that the basis of the EU policy regarding GM foods is a precautionary principle (and fact) that consumers should know what they are eating. In case of uncertainty it is always better to take precaution. Similarly, the EU believes that there is an element of uncertainty in the case of GM foods, regarding their effect on human health and the environment. Therefore, it is advisable to be cautious while dealing with the same. As per the Regulation on GM Food and Feeds of the EU: "For non pre-packaged products offered to the final consumer the words 'this product contains genetically modified organisms' shall appear on or in connection with the display of the product." Furthermore, the Regulation on GM Food and Feed states that the granting of authorization will not lessen the general civil and criminal liability of any food operator in respect of the food concerned.[116] The UK Agricultural and Environment Biotechnology Commission and the UK Department of Trade and Industry have joined hands in this regard. The UK government initiated a public debate on GM crop commercialization for mapping public opinion to ensure that the public is given a fundamental right to get their voice heard in case of GM foods.

---

[112] Regulation (EC) No: 1829/2003 on genetically modified food and feed.

[113] Regulation on Traceability and Labeling, OJ L 268/24, 2003.

[114] Directive 2001/18/EC.

[115] See Evenson, Robert E. and Terri Raney (eds). 2007. *The Political Economy of Genetically Modified Foods*, p. 294. UK: Elgar Reference Collection.

[116] See Article 7.7 of the Regulation on GM food and Feeds of the EU.

Indian Experience with GM Food

With the ever increasing population in India, ensuring the right to food and food security is a major concern. If biotechnology is able to help the nation in providing the right to food and food security, it would be a great achievement from the Indian perspective. India is on shaky grounds when it comes to GM foods, due to public oppositions and apprehensions and also for lack of effective regulatory structure. The same could be witnessed in the fiasco of Bt brinjal, where, in 2011, the government itself wanted to permit marketing approvals for Bt brinjal. But these were withdrawn, and an indefinite moratorium was imposed on the crop in 2012. Stakeholders from different quarters strongly emphasized on having a proper and trustworthy regulatory structure in place before getting ready for clinical trials and marketing approvals for GM foods in India. Food security has been a major developmental objective in India; even though it achieved self-sufficiency in foodgrains in the 1970s and has sustained it since then.

Ensuring food security is not completely dependent on GM crops only, although field trials on Bt rice and other food crops were conducted in India, projecting them a step forward toward ensuring food security. In 2004, Bt rice field trials were conducted in Jharkhand,[117] which could not give credible and substantial results for the government to clearly approve Bt rice or other Bt food crops in India. This could be due to number of other factors including increasing cultivation lands, water resources, drip irrigation, crop change practices, monsoon, and regular and non-Bt on GM innovations, which would contribute to the development of agricultural sector in terms of increasing yield, reducing the cost of production, and ensuring food security. Bt brinjal[118] is the first GM food crop in India to reach the stage of approval for commercialization. It has raised much controversy in India, especially since it is a food crop.

[117] Sahai, Suman. 2014. 'We do not have the competence to play around with GM foods', *The Times of India* (Kolkata edn), 2 November, available at http://timesofindia.indiatimes.com/home/sunday-times/all-that-matters/We-do-not-have-the-competence-to-play-around-with-GM-foods-Suman-Sahai/articleshow/45009714.cms (last accessed 3 November 2014).

[118] The transgenic brinjal is created by inserting a gene (Cry1Ac) from the soil bacteria Bt. It is said to make the brinjal plant pest and insect resistant.

The contentious points revolving around Bt brinjal are common to any GM crop.

The promoters of Bt brinjal assert that Bt brinjal will increase crop yield, decrease the cost for farmers since it is pest resistant, and it has no health hazard. On the other hand, the opposition contends that there could be adverse effects on human health and biosafety.[119] India is the second largest producer of brinjal in the world: approximately 2,500 varieties of brinjal are grown in India. The public in India wants to know the reason for the need to go for a Bt variety of brinjal, as neither there is a scarcity of brinjal nor there is a lack of variety. Bt brinjal has been produced and marketed by Mahyco, India, with Monsanto of the US, being a minor partner. There are other two public-funded institutions involved in the production of Bt brinjal, namely, the University of Agricultural Sciences, Dharwad, Karnataka, and the Tamil Nadu Agricultural University.

A bill, Protection and Utilisation of Public Funded Intellectual Property (PFIP) Bill, 2008, was presented before the Parliament but got lapsed. There is a growing consensus that qualified people with experience in the respective field should be involved in the process of considering the suitability of Bt products, including their field trials, clinical trials, and marketing approvals. Vandana Shiva,[120] Suman Sahai,[121] Jayashree Watal,[122] and other such popular social activists raised strong

[119] Ministry of Environment and Forests (MoEF) and Centre for Environment Education (CEE). 2010. *National Consultations on Bt Brinjal: A Primer on Concerns, Issues and Prospects*, Nehru Foundation for Development, Ahmedabad, available at http://www.moef.nic.in/downloads/public-information/Bt%20Brinjal%20Primer.pdf (last accessed 1 March 2016).

[120] Sahai, Suman, Vandana Shiva, Aruna Rodrigues, and Kavita Kurugant. 2014. 'India's Sovereignty, Security and Freedom at Risk: Is the IB being used by foreign corporations to take over India's vital seed sector?' available at http://vandanashiva.com/?p=50 (last accessed 21 June 2016).

[121] Sahai. 2014. 'We do not have the competence to play around with GM foods'.

[122] Watal, Jayashree. 2004. 'Intellectual Property Rights in Agriculture', in Merlinda D. Ingco and L. Alan Winter (eds), *Agriculture and the New Trade Agenda: Creating a Global Trading Environment and Development*, pp. 401–27. New York: Cambridge University Press.

objections against Bt brinjal and alleged that organic brinjal is more fruitful, safe, and healthy, when compared with Bt brinjal and there is no need to promote Bt brinjal to appease multinational companies and their profit motives. During the period of 2009–12, atmosphere was not in favor of approving Bt brinjal in India, the then Minister for Environment, Jairam Ramesh, was instrumental in collecting opinions from stake holders across the country, including Hyderabad, Nagpur, Mumbai, Ahmadabad, etc.[123]

The Indian experience with Bt crops has not been good: Bt cotton turned out to be catastrophic, leading to farmer suicides in Maharashtra and Gujarat. The last two decades, that is, 1996–2016, have been a learning experience for the farmers in these regions. With such disastrous experience, the approach should have been a cautious "wait and watch" with decisions being taken after consulting all the stakeholders. There should not be any urgency, since India does not require Bt brinjal. If the intention is to promote innovation and creation, it should not be at the cost of health, environment, and other social concerns of the society. The economic motives of a few people should not be prioritized above all the human concerns in this regard. Even the success of biotechnology is not completely established, in terms of whether GM food is completely safe from every respect or not. Therefore, things should not be done in hurry with reference to biotechnology, its techniques, and products.

The Supreme Court of India on GM Foods

The Supreme Court of India got an opportunity to look into the matter of GM crops, GM foods, and food security in *Aruna Rodrigues and Ors v. Union of India*.[124] The petitioner approached the Supreme Court against the decision of the Government of India in allowing field trials of GM crops and GM foods. In response, the Supreme Court constituted a technical expert committee to look into the issue at hand. The committee recommended that there should be a moratorium on field trials

---

[123] See http://www.genet-info.org/information-services/bt-eggplants-brinjal-in-india/page/3.html (last accessed 21 June 2016).

[124] Writ Petition (Civil) No. 260 of 2005, *Aruna Rodrigues and Ors v. Union of India*.

for Bt in food crops (those that were directly used as food) intended for commercialization (not research) until there is more definitive information from a sufficient number of studies about the long-term safety of Bt in food crops. It was observed that the single largest number of applications for field trials to the Genetic Engineering Advisory Committee (GEAC) was for Bt transgenics, basically including food crops such as rice and a range of vegetables.[125] Regarding the nature of tests for Bt in food crops, the committee was of the view that there should be test to check the chronic toxicity of Bt transgenics. It may also be noted that by far the largest deployment of transgenics worldwide is in soybean, corn, cotton, and canola, all of which are used primarily for oil or feed. Nowhere are Bt transgenics being widely consumed in large amounts for any major food crop that is directly used for human consumption. The committee could not find any compelling reason for India to be the first to do so. It may be noted that the moratorium on Bt crops is now indefinite, the cap of 10 years being removed, as imposed by the Ministry of Environment and Forests (MoEF) on Bt brinjal, until all the safety conditions are met. These safety conditions deal with two broad aspects: (a) regulatory independence and expertise, and (b) the requirement of comprehensive risk assessment and hazard identification protocols to ensure biosafety.[126] It is very pertinent to note here that the Cartagena Protocol on Biosafety, International Plant Protection Convention, and the WHO Codex Alimentarius Commission attempt to ensure safety in the use of biological resources, plant genetic resources, and also food.[127] It is clear that biosafety has been clearly articulated and is the thrust of the report of the technical expert committee appointed by the Supreme Court of India. The committee also recommends reexamination of the safety data of the approved applications to ensure that all the biosafety issues have been addressed, including long-term multigenerational feeding studies for chronic toxicity.

[125] Out of the 91 applications referred to GEAC, 44 were on field trials of rice, vegetable, and other such consumable good crops.

[126] View, generally, http://www.gmwatch.org/index.php/news (last accessed 10 August 2014).

[127] See Francioni and Scovazzi (eds). 2006. *Biotechnology and International Law*, p. 264.

Labeling of GM Food: Right to Know Bt Crops and Food Security

Labeling of GM products may be desirable for several reasons, including both health and environmental concerns. Labels can provide information about allergens, toxins,[128] change in cooking characteristics, or nutritional content of food, and the ethical or environmental concerns, and can help to prevent deceptive practices. Consumers have every right to know what they have been offered, what options they have, and what they are eating. In case there is a food safety concern related to toxins, a labeling requirement would be covered by the Agreement on the Application of Sanitary and Phytosanitary Measures (SPS agreement); if there is a labeling requirement due to concerns related to the nutritional characteristics, not related to food safety, it would be covered by the TBT agreement. The TBT agreement talks about labeling of GM food released for human consumption. According to scientific evidence, certain potential risks are associated with the GM food and a regulatory framework was required to address this pertinent issue. However, nations are divided on the adoption of policy framework that could enable the authentication and labeling of GM food. The policies of the US and the EU represent different approaches toward the regulation of GM food and its labeling. In 1992, the Food and Drug Administration (FDA), in a policy document outlining the US policy, recommended voluntary labeling of GM food. The policy opposes mandatory labeling of GM food in general.[129] Further in 2002, the modified policy regarding GM food reiterated its opposition to mandatory labeling. There is no mandatory labeling of GM foods in the US. It is left to the producer of the GM food to either label the food or not. Canada also follows the US policy in not prescribing mandatory labeling of GM foods.

Unlike the US where there is no requirement of premarket approvals for GM crops, in Europe there was moratorium on environmental

---

[128] Naidu, David. 2009. *Biotechnology and Nanotechnology: Regulation under Environmental, Health and Safety Laws*, p. 243. London: Oxford University Press.

[129] Evanson and Raney. 2007. *The Political Economy of Genetically Modified Foods*, p. 53.

release of GM crops till 2001. The European Parliament, in 2001, voted for stricter rules regarding the regulation of GMOs including mandatory labeling of GM food, while obligating the makers of GM food to ensure compliance of the same.[130] Similarly countries such as Brazil, Thailand, Japan, South Korea, Taiwan, Australia, and New Zealand have had moratorium on GM crops and prescribe mandatory labeling of GM food.[131] Countries such as India and Norway have been supporting mandatory and comprehensive labeling not only on safety and suitability grounds but also on ethical grounds. So, the international community is divided on the mandatory labeling of GMOs. While certain international groups and organizations, such as Greenpeace International, Friends of the Earth, and the Consumer Union, are voting for mandatory labeling of GM food, the Council for Biotechnology Information and US FDA are against the same. The recommendations of the Codex Alimentarius are also in favor of comprehensive labeling of GM food.[132] India, being a member country of the World Trade Organization (WTO), also requires labeling as per the TBT agreement and also to warn consumer from any health hazards that GM food may cause as per the SPS agreement. However, no major GM food crop is given full marketing permission in India; so as of now from the point of view of consumers' rights, labeling of GM food has not yet become a major concern in India, especially for the purpose that it serves: the right to know the contents of the food.

GM Foods and Consumer Rights

In the contemporary times, when public awareness is increasing day by day, consumers are at the demanding end, not at the receiving end. The producer is supposed to feed the needs of the consumers to their satisfaction in accordance with the popular demand. The consumers would like to know the details of the product they are purchasing including how it is

---

[130]   See Official Journal of European Communities, 2000, L006/13-17.

[131]   Evanson, R. and V. Santaniello (eds). 2004. *Regulation of Agricultural Biotechnology*, p. 54. Wallingford, UK: CABI Publishing.

[132]   See 'Understanding of Codex Alimentarius,' available at www. Fao.org/docrep/W9114E/W9114E00.htm (last accessed 1 March 2016).

made and how it would perform.[133] Even though the labeling of GM food has been made mandatory from 2013 in India, the move is not enough to provide for people's right to know, ethical and religious concerns, amongst others. First, because more than 90 percent of food in India is unprocessed, unpackaged, and comes from the unorganized sector. Second, there is no clarity on threshold for the presence of GM ingredients,[134] how the government intends to monitor the mechanism, and whether the new law is applicable to both primary and processed foods. Third, many administrative problems have not been addressed. Testing methods have not been revived since several years. Due to the impending Biotechnology Regulatory Authority of India Bill, 2013, there is no regime for controlling imports and custom authorities are still not equipped to distinguish a consignment of GMO from that of a non-GMO.[135]

Presently, procedures for evaluating food safety mentioned in the guidelines are closer to studies carried out for chemical pesticides. Scientific experts recommend that food safety guidelines for transgenic crops should be similar to those for food additives. The guidelines provided by the Department of Biotechnology (DBT), Ministry of Science and Technology in India for allergenicity need to be modified and strengthened as the current ones are outdated and protocols do not provide full information. The government needs to provide for product liability laws to enforce corporate responsibility. It should be made mandatory for the industry to label transgenic seeds that are sold to farmers. Moreover, even though many laboratories in India follow good practices, there is a pressing need to follow the Good Clinical Laboratories Practice[136] developed by the WHO in its entirety.[137]

[133] Sreenivasulu, N.S. 2014. 'Consumer Interest in the Regulation of Trademark Propriety', *International Journal of Consumer Law*, 1(1): 79–88.

[134] Nandi, Jayashree. 2012. '"GM" Label on Packaged Food Soon', *The Times of India*, 18 June.

[135] Khanna, S.R. 2001. 'Genetically Modified Foods: The Consumer's Concern', in *Relevance of Genetically Modified Plants in Indian Agriculture*, p. 173. New Delhi: Tata Institute of Renewable Energy.

[136] Selvakumar, R. 2010. 'Good Laboratory Practices', *Indian Journal of Clinical Biochemistry*, 25(3): 221–4.

[137] See World Health Organization. 2009. *Laboratory Biosafety Manual*, Geneva, available at http://www.who.int/tdr/publications/documents/gclp-web.pdf?ua=1 (last accessed 2 March 2016).

The Food Safety and Standards Act, 2006 (FSS Act), is the primary legislation amongst others to ensure that the GM food is fit for human consumption. The Ministry of Consumer Affairs, through a gazette notification,[138] has made it mandatory to label food containing GM organisms from 1 January 2013. India also strives to follow the guidelines of the International Codex Alimentarius for food quality and standards.[139] The objective of the Food Safety and Standards Authority of India (FSSAI) is to ensure safety of the food produced, supplied, and consumed in India, including GM food. The National Food Security Act, 2014, which intends to provide for right to food to all is perhaps a very significant legislation. The legislation acknowledges the contribution of biotechnology and its innovations in providing high yield and quality of food products. Foods produced through biotechnology or otherwise should fulfill standards set under the FSS Act. While ensuring food security, the quality of the food needs to be ensured by adopting the prescribed standards set by the FSSAI.

Acceptability of GM Foods

Several of the undertaken surveys have proved that there is a serious lack of information and knowledge about biotechnology in the country.[140]

[138] Ministry of Consumer Affairs, Government of India through a gazette notification on 5 June 2012 made it mandatory to label GM foods with effect from 1 January, 2013. See http://www.indiaenvironmentportal.org.in/category/3840/thesaurus/ministry-of-consumer-affairs-food-and-public-distribution/ (last accessed 21 June 2016).

[139] See Food and Agricultural Organization of United Nations (FAO). 2009. "User's Manual on Codex—India: A Contemporary Approach to Food Quality and Safety Standards," FAO User's Manual, TCP/IND/0067, Ministry of Health and Family Welfare Government of India, New Delhi.

[140] Knight, John and Amit Paradkar. 2008. 'Acceptance of Genetically Modified Food in India: Perspectives of Gatekeepers', *British Food Journal*, 10(110): 1019–33. See also Deodhar, Satish Y., Sankar Ganesh, and Wen S. Chern. 2007. 'Emerging Markets for GM Foods: An Indian Perspective on Consumer Understanding and Willingness to Pay', Indian Institute of Management Ahmedabad, India, W.P. No.2007-06-08 June 2007, available at http://www.iimahd.ernet.in/publications/data/2007-06-08Deodhar.pdf (last accessed 2 March 2016).

Since public acceptance forms a major role in the making and implementing of laws in India, it is imperative that the confidence of the public is won by spreading awareness. Given that perceptions about impact and risks along with costs and benefits of biotechnology are so diverse, it is not surprising that consensus remains elusive when discussions turn to policy. Much of the hindrance in policymaking and implementation occurs due to lack of public acceptance. It has been seen that the reason for weak public acceptance in India is largely due to half-truths and deliberate propaganda from interested groups.[141]

The blanket moratorium on field trials of GM crops, including Bt brinjal, was imposed in 2012 aftermath of serious public objections and apprehensions. The government went on to note the public perceptions regarding Bt foods by conducting stakeholders' meetings across country, which found that there is a need for an immediate ban on GM foods and for deeper investigations into the issue of GM crops and their safety. The parliamentary standing committee was also constituted in this backdrop, which backed the public perceptions against GM crops in India. Research institutions should therefore make their research public and communicate it to public and policymakers to boost their confidence in biotechnology. There has to be an increase in the number of bioinformatics networks currently operating in India to cover district-level populations as well. Local awareness campaigns coupled with enhanced regulatory transparency will prove fruitful in gaining public trust.

It must also be understood that throughout human history, technological developments such as electricity, automobiles, air travel, etc., all have posed certain risks and threats. However, humankind has evolved through such risks and it has not prevented them from benefiting by such advancements. The fate of biotechnology is no different. In fact, it is interesting to note that GM products have been subjected to rigorous testing procedures whereas there was no testing mechanism adopted for conventional methods of agriculture. However, this is not to say that there are no unforeseen risks with GM products. Most of the issues

---

[141] Sreenivasulu, N.S. and Debashu Chettary. 2016. 'Agricultural Biotechnology: Environmental Law and Other Policy Considerations', *Environmental Law Practice Review*, (I): 25–39.

can be addressed through appropriate research rather than emotional debates or militant activism. As philosopher John Rawls puts it, while deciding on policy issues, we must put the veil of ignorance in front of us according to which we must be blind to the personal impacts of the policy.[142] However, it is to be noted that in India certain GM foods such as GM soya oil and corn syrup are already in public consumption. Scholars argue that there are doubts about the safety of such GM foods already in public consumption and tests for toxicity, allergenicity,[143] or any other health impacts, should be conducted. There cannot be mass or blind dependence on biotechnology and GM crops.

The declared primary objective in international development policy discourse is the reduction and elimination of poverty.[144] Food security is a problem not only of production but of distribution and access or purchasing power. Today, India's paradox of overflowing godowns and rotting grains, with 320 million people going hungry, is well known. The world over and in India, most of the hungry people are ironically participating in the food production process. Clearly, hunger is a more multifaceted problem than what can be fixed by using a particular seed or a cocktail of chemicals. Therefore, the government's attention is sought to address the issue of food security in a holistic manner, taking into account the issue in its totality rather than looking for shortcuts and getting distracted by red herrings such as GM crops and pesticides. It has to be remembered that although biotechnology is an option, it is not the only way or method to ensure the future of the agriculture. In the light of varied voices from different stakeholders in the policymaking, the Government of India should adopt the approach of detailed consultation and investigation into the various stakes before taking a final call on pushing field trials and commercial cultivation of Bt and GM crops in India. There are a number of non-Bt or non-GM innovations, breakthroughs, and

---

[142] See generally, Rawls, John. 1971. *Theory of Justice*. Cambridge, MA: Harvard University Press.

[143] See also Sahai. 2014. 'We do not have the competence to play around with GM foods'.

[144] FAO. 2003. 'Food Security: Concepts & Measurement,' available at http://www.fao.org/docrep/005/y4671e/y4671e06.htm (last accessed 1 December 2012). Reproduced in "Food Security" Policy Brief, FAO, June 2006, Issue 2.

breeding successes including tissue culture and organic farming, which should also be promoted for sustaining the development of agriculture sector and for ensuring food security.

Biotechnology and the Right to Health

Application of biotechnology in the field of medicine aims to provide for the right to health. The novel innovations in biotechnology such as biopharmaceuticals, biological cures through genetic engineering, cure of hereditary diseases through gene therapy, and other such innovations and breakthroughs in biotechnological R&D hold promise for improving public health.[145] At the same time, biotechnology could also affect the public health sector negatively. It is argued that patents on biotechnologically produced healthcare products would push the control of the much-needed healthcare products into private hands, which might make them unaffordable. Such patents may block further R&D in the field.

The "right to health" has four interrelated and essential dimensions:[146] availability, accessibility, quality, and acceptability. The right is guaranteed under the Indian constitution and under a couple of international conventions. The UDHR[147] proclaims that every human being has the right to share scientific advancement and its benefits freely. It is quite relevant in the case of biotechnology due to its promising advancements to cater to the food, health, and other such related needs of the community. Article 12 of the ICESCR requires member parties to recognize the right of everyone to access the highest attainable standard of physical and mental health.[148] It enumerates nonexhaustive obligations of states

---

[145] See Sreenivasulu. 2008. *Biotechnology and Patent Law.*

[146] Smith, Alyna C. 2007. 'Intellectual Property Rights and the Right to Health: Considering the Case of Access to Medicines', in Christian Lenk, Nil. S. Hope, and Roberto Andorno (eds), *Ethics and Laws of Intellectual Property – Current Problems in Politics, Science and Technology*, pp. 47–72. Hampshire, England: Ashgate Publishers.

[147] UDHR was adopted in the year 1948, and a number of nations including the US, European countries, and India are parties to the Declaration.

[148] Varghese, John and N.S. Sreenivasulu, N.S. 2016. 'Gene Patents and Access to Health', in N.S. Sreenivasulu (ed.), *Human Rights and Development*, pp. 185–200. Bloomington, US: Penguin-Partridge Publications.

parties, which include the creation of conditions that would assure medical services and medical attention to all. The Universal Declaration on Human Genome and Human Rights states that the application of research including applications in biology, genetics, and medicine, concerning the human genome, must offer relief from suffering and improve the health of individuals and humankind as a whole. Since India is a signatory, it is bound to fulfill its obligations enumerated by ICESCR and UDHR and to facilitate the "right to health" to its citizens.

In the Constitution of India, Article 21 guarantees protection for life and personal liberty by providing that no person would be deprived of his life or personal liberty except according to the procedure established by law. This right has been interpreted to include the right to health by the law courts. The state is obligated to protect and improve public health as mentioned under the Directive Principles of State Policy under the constitution:[149] it should introduce and implement various international instruments that are consistent with the fundamental rights and in harmony with its spirit including the fundamental right to health. The right to health is guaranteed under the constitution as a component of right to life. In addition, the state is obligated to create an environment to ensure the right to health of the people. In *M.K. Sharma v. Bharat Electronics Ltd*[150] it was viewed that Article 21 of the Constitution has been held to include "right to health."

The Supreme Court of India has expanded the repertoire of Article 21 so the life becomes meaningful instead of a mere vegetative existence.[151] It was stated that the maintenance and improvement of public health should rank high, as these are indispensable to the very physical existence of the community, and the composition of the society as envisaged by the constitution depends on the betterment of these factors. Furthermore, in *Vishaka v. State of Rajasthan*,[152] the Supreme Court of India finds that under the constitution the state is obligated to respect international law and treaty obligations. This decision was made

[149] See *People's Union for Democratic Rights v. Union of India* 1982 AIR SC 1473: 1983 SCR (1) 456.

[150] AIR 1987 SC 1792.

[151] *Vincent Panikurlangara v. Union of India* 1987 AIR SC 990: 1987 SCR (2) 468.

[152] AIR 1997 SC 3011.

in the light of international conventions recognizing right to health and medical assistance. However, in *M.C Mehta* v. *Union of India*[153] the Supreme Court of India held that Articles 47 and 48A individually and collectively caste a duty on the state to secure the health of the people, improve public health, and protect and improve the environment. Hence, the Indian policymakers must keep in mind the Directive Principles, as articulated under Articles 47 and 48A, to ensure that every regulation should be in consonance with the state goals under the constitution. As a matter of fact, the state is obligated to respect the human rights in general and the right to health in particular, which are very much recognized under the International Convention on Economic, Social and Cultural Rights adopted in 1966. In other words, it is the duty of the state to apply these principles while making laws. Policymakers and courts have often debated and discussed these various dimensions of right to health in the making of law and as well in the process of adjudication of justice. In the sphere of biotechnology, the right to health is poised to be guaranteed through the application of genetic technology and genetic engineering. However, certain negative implications of genetic technology and its innovations were predicted on the right to health. In particular, innovations such as genes and patenting of the same may block the valuable innovations that could cater to the health needs in few hands.

Genetic Research and Right to Health

Of the four components of "right to health", the component "availability" requires the presence of adequate healthcare facilities so as to meet the needs of the population. Such facilities must also be "accessible" to all the sections of the population without any physical and economic discrimination. It is argued that gene patents on the various gene technology innovations act as gatekeeper patents, thus infringing upon both these facets of "right to health." These patents claim underlying fundamental information about genetic behavior that is pertinent for future, down-stream research. The very nature of gatekeeper patents allow them to entail that all the uses of the gene under the patent, including gene therapy and pharmacological modulation of the gene, must go through

[153] JT 2002 (3) SC 527; MANU/SC/0254/2002.

original gene patent before an invention using that gene can be practiced.[154] Thus, the patentee ultimately "controls" all work deriving from
the use of that gene.

The United States Patent and Trademark Office (USPTO), for
instance, granted patent to Human Genome Sciences (HGS), which
claimed rights to a gene, the precise function of which was initially
unknown and the utility of which was asserted to be a research reagent
or material for diagnostics.[155] When other researchers subsequently discovered that the DNA sequence actually coded for CCR5 receptor,[156]
which is the "docking receptor" used by the HIV virus to infect a cell, it
was widely feared that this patent would have a *blocking* effect on AIDS
research. Since HGS had issued several licenses for research into new
drugs and did not prevent academicians from undertaking unlicensed
research on CCR5, this fear became unfounded.[157] But had HGS decided
otherwise, it would have had a devastating effect on AIDS research.

Products that do not have prima facie relation to the gene in question
and are researched independently of the gene may also require licensing from the patentee before that product can be used. For example,
an inventor could create a drug that itself will not infringe any other
product patent but will infringe the gene patent if that drug modulates
the patented gene. The proliferation of gene patents, including multiple
patents on various research tools, can necessitate negotiating multiple
licenses when developing a single product or process. Such "patent
thickets"[158] have the potential to raise the transaction costs of doing

---

[154]   Barton, John H. 1997. 'Patents and Antitrust: A Rethinking in Light of
Patent Breadth and Sequential Innovation', *Antitrust L.J.* (65): 455–61.

[155]   Varghese and Sreenivasulu. 2015. 'Gene Patents, and Access to Health',
pp. 185–200.

[156]   CCR5 (chemokine [C-C motif] receptor 5) is a protein that, in humans,
is encoded by the CCR5 gene. They are predominantly expressed in T cells and
hence play a vital role in inflammatory responses to infection.

[157]   'Genetic Inventions, Intellectual Property Rights and Licensing Practices—Evidence and Policies', Organisation for Economic Co-operation and
Development (OECD), available at http://www.oecd.org/dataoecd/42/21/
2491084.pdf (last accessed 1 March 2016).

[158]   The term "patent thicket" has been coined to characterize a technological
field where multiple rights owned by multiple actors may impede R&D owing
to the difficulty or cost of assembling the necessary rights.

research and possibly the ultimate cost of products owing to stacking of royalties. Research tool patents (for example, patents on markers, assays, receptors, and transgenic animals) claim products "identified by" the patented tool or method. If such a claim is granted, patent owners can demand royalties on the sale of a product found with the help of their research tool. Since many different patented research tools must be used in the development of a drug, such reach-through claims increase royalty stacking.

The DuPont–Cre-lox controversy[159] is the archetypal example of the aforementioned concerns. Cre-lox is a gene-splicing tool patented by Harvard University. It was exclusively licensed to DuPont Pharmaceutical Co. but the company asked researchers to sign an agreement that would limit their ability to use and share the Cre-lox technique and subject their articles to prepublication review by the company. DuPont wanted commercial rights over future inventions that might arise from experiments using transgenic animals, which involved the use of Cre-lox splicing tool (that is, reach-through rights). Some prominent institutions, including the National Institutes of Health (NIH), refused to sign the agreement, claiming that this created obstacles to biomedical research. Finally, a memorandum of understanding was signed between NIH and DuPont (and separate agreements with academic laboratories), which simplified access conditions for the US public sector to this patented research tool.[160]

There has also been concern that the rapid proliferation of gene patents will increase commercial uncertainty owing to possible interdependence between granted patents. For example, if different patents are granted for inventions that claim, respectively, a partial gene sequence, the full-length complementary DNA (cDNA) or gene, and the encoded protein, it is unclear which title holder will be able to prevent the others from using his or her invention. It is thus evident that the patenting of genes increases cost of R&D and of drugs and medicine, and affects the right to health.

[159] See George, Mathew P. and Akanksha Kaushik. 2010. 'Gene Patents and Right to Health', *NUJS Law Review*, (3): 323–35.

[160] Varghese and Sreenivasulu. 2016. 'Gene Patents, and Access to Health,' pp. 185–200.

The human genome sequencing project has brought about many potential applications for therapeutic uses. Gene therapy was reported to correct sickle-cell anemia, an inherited disorder caused by a mutation in the beta-globin gene that causes individuals to manufacture abnormal hemoglobin. There are at least 5,000 genetic diseases that do not have any treatment. Genetic sequences can determine the propensity of an individual to contract a disease and cure the disease through gene therapy or targeted pharmacological modulation of the gene itself or the protein that the gene codes for.[161] For example, detection of the mutations alerts physicians to the probability of the individual in contracting breast, ovarian, and other types of cancer. Patients with these mutations may be referred for genetic counseling to evaluate the risk of surgical intervention or explore the scope of gene therapy.[162] However, granting of patents for genes can deprive a large section of the masses of the immense benefits that such improved healthcare can provide. Thus, it is believed that patenting of genes infringes the right to health and access to quality healthcare at an affordable price. Gene therapy, that is, correction of genetic defects within the cell genome, has varying levels of success dependent upon the model system being studied.[163] Such therapy requires genetic sequence information to explore the scope of mutation and correct the defect in case the sequence is mutated. Therefore, all therapy products used in the gene therapy must flow from the original genetic information source.

Most of the drugs work only on a certain percentage of patients who use them.[164] Genetic testing can help distinguish those patients for whom a drug will work from those for whom it will not. India is bound to fulfill

[161] Malinowski, Michael J. and Maureen A. O'Rourke. 1996. 'A False Start? The Impact of Federal Policy on the Genotechnology Industry', *Yale J. on Reg.* 13: 163, 165–7.

[162] For example, research revealed that mutations in the genetic sequences of two genes (*BRCA-1* and *BRCA-2*) can be linked to hereditary forms of breast cancer.

[163] Gene therapy vectors are host organisms, usually a virus, that carry the DNA to be inserted or substituted directly to the target cells of the recipient organism.

[164] Roses, Allen D. 2000. 'Pharmacogenomics and the Practice of Medicine', *Nature*, 405: 857.

its obligations under international law to facilitate the enjoyment of "right to health" by its citizens. The legal uncertainty surrounding the patenting of genes is poised to be unacceptable considering the rapid growth of biotechnology in India. The lawmakers will have to work a policy in this regard to ensure the right to health of the people and, at the same time, use biotechnology for enhancing the quality of healthcare and the right to health.

7 Biotechnology Regulation in India

egulation of biotechnology is at a nascent stage in India. The issue of regulation of biotechnology has been contentious regarding the mode and method of application of law in the field of biotechnology. Perhaps, regulation of any technology should be done with a holistic approach to ensure optimal utilization of such technology toward sustainable development. Attempts to provide a regulatory mechanism for biotechnology in India started about 26 years ago, with the introduction of the Rules for the Manufacture, Use/Import/Export and Storage of Hazardous Micro Organisms/Genetically Engineered Organisms or Cells by the Government of India in 1989. Since then, a number of guidelines, rules, regulations, and legislations have been instituted to ensure environmental safety in the use of genetically modified organisms (GMOs); provide regulatory approvals for dealing with genetically engineered (GE) entities; promote research in biotechnology, including the provision of patent rights on novel biotechnology innovations; protect and maintain biological diversity; improve the food and health needs of the country; and set up a national-level authority on biotechnology regulation in India.

In order to ensure optimal utilization of biotechnology toward sustainable development, the laws and policies should be in compliance with the law of the land—the Constitution of India and the socioeconomic goals set out in the Directive Principles of State Policy prescribed under the constitution. At the same time, the fundamental rights of the people of India are also required to be safeguarded in the most dynamic way by any policy that intends to regulate biotechnology. In this respect, although GMOs have not been specifically dealt with under the domestic laws of India, there have been piecemeal attempts to address the aforementioned aspects of compliance by way of including guidelines, regulations, legislations, and measures to establish an institutional mechanism. These include the following:

- Rules for the Manufacture, Use/Import/Export and Storage of Hazardous Micro Organisms/Genetically Engineered Organisms or Cells, 1989
- Guidelines such as:
  - Recombinant DNA Safety Guidelines, 1994 (adopted in 1990)
  - ICMR (Indian Council of Medical Research) Guidelines, 2000
  - Revised Guidelines for Research in Transgenic Plants and Guidelines for Toxicity and Allergen City Evaluation, 1998
  - Guidelines for Stem Cell Issued By the ICMR in Association with Department of Biotechnology, Government of India, 2012
  - Guidelines for the Conduct of Confined Field Trials of Transgenic Plants, 2008
  - Guidelines for the Safety Assessment of GM Foods, 2008
  - Protocol for Safety Assessment of Genetically Engineered Plants and Crops, 2008
- Legislative efforts such as:
  - Protection of Plant Varieties and Farmers' Rights Act, 2001
  - Patents (as amended in 2005) Act
  - The Biodiversity Act, 2002

- DNA Profiling Bill, 2015
- Biotechnology Regulatory Authority of India Bill, 2013[1]
- A few committees have been constituted under the 1989 rules and the Recombinant DNA (rDNA) safety guidelines, 1990 at the institutional level:
  - Genetic Engineering Approval Committee
  - Recombinant DNA advisory committee
  - Review committee on genetic manipulation
  - The Institutional Bio-safety Committee
- Further there are also state biotechnology coordination committees and district-level committees.
- The Indian parliament appointed the parliamentary committee which functioned during 2009–12 and the Supreme Court of India appointed the Technical Advisory Committee in July 2012.

The following sections discuss each of these legal and policy initiatives in the direction of regulation of biotechnology in India. They also analyze the institutional mechanism that has been established in India with reference to the regulation of biotechnology as well as its innovation and use.

Rules for the Manufacture, Use/Import/Export and Storage of Hazardous Microorganisms/ Genetically Engineered Organisms or Cells, 1989

These guidelines were notified by the Ministry of Environment, Forest and Climate Change, Government of India on 5 December 1989, under the provisions of the Environment (Protection) Act, 1986.[2] The rules were made to protect human health in relation with the application of gene technology and microorganisms, and as well to contain possible hazards to the environment due to the release of GMOs. These rules are applicable to:

---

[1] The bill that was introduced in the year 2013, and is still pending before the parliament.

[2] Through their Notification No. 621 in *The Gazette of India*, Government of India.

- manufacture, import, and storage of microorganisms and gene technological products;
- GE organisms/microorganisms and cells and correspondingly to any substance and food products of which such cells, organisms, or tissues form a part; and
- new gene technologies in addition to cell hybridization and genetic engineering.

The sale, offers for sale, storage for the purpose of sale and offers, and any kind of handling with or without a consideration of GMOs would fall within the purview of the rules. Moreover, exportation and importation of GE cells or organisms, production, manufacturing, processing, storage, import, drawing-off, packaging, and repacking of GE products would come under the scanner of these rules. Besides, production and manufacture of drugs, pharmaceuticals, food items, distilleries, and tanneries, which make use of microorganisms or GE microorganisms in one way or the other, would also be regulated under the rules. In addition, to ensure clarity in dealings with biotechnology, the rules define the terms "biotechnology," "cell hybridization," "gene technology," "genetic engineering," and "microorganism."[3] To ensure proper regulation, microorganisms or GE products have been broadly classified as animal pathogens and plant pets. Further, microorganisms laid down in the schedule are divided into six major heads: bacterial agents, fungal agents, parasitic agents, viral agents, rickettsial agents, chlamydial agents, and special category.

For the regulation of biotechnology and its use, the rules propose institutional mechanisms. The following competent authorities have been established under the rules in this regard:[4]

- Recombinant DNA Advisory Committee (RDAC)
- Institutional Bio-safety Committee (IBSC)
- Review Committee on Genetic Manipulation (RCGM)
- Genetic Engineering Approval Committee (GEAC)

---

[3] Section 3 of the 1989 rules.
[4] Section 3 of the 1989 rules.

Recombinant DNA Safety Guidelines, 1994

The Department of Biotechnology, Ministry of Science and Technology, Government of India, in January 1990, issued a set of guidelines called rDNA Safety Guidelines, revised in 1994. These guidelines cover a plethora of issues pertaining to rDNA technology ranging from a checklist of risk assessment factors on environmental release of genetically manipulated organisms, import and shipment, containment facilities and biosafety[5] practices, etc. Areas of research involve GE organisms; genetic transformation of plants; rDNA technology in vaccine and diagnostics development; large-scale production and deliberate and accidental release of organisms, plants, animals, and products derived by rDNA technology into the environment; as well as the import and shipment of GMOs or transgenics for laboratory research and their large-scale use.

The guidelines can be considered a step ahead from the 1989 rules on the use of microorganisms issued by the Ministry of Environment, Forest and Climate Change, acknowledging this institutional mechanism through the creation of authorities and committees to ensure biosafety. The committees under the rules have been further empowered under the guidelines with specificity, as discussed further under "Institutional Mechanism for Biotechnology Regulation in India." They provide for good laboratory practices to help the scientific community in properly conducting and maintaining their experiments with microorganisms and DNA, which could ensure safety of the environment and the health. The summary of recommended biosafety levels are also provided under the guidelines along with containment conditions for plant experiments and containment approaches for large-scale industrial applications that would help and assist scientific experiments to be in tune with the safety regulations.

ICMR Guidelines, 2000

ICMR issued guidelines governing the medical research, titled "Ethical Guidelines for Biomedical Research on Involving Human Participants",

---

[5] See generally, Evenson, Robert E. and Terri Raney (eds). 2007. *The Political Economy of Genetically Modified Foods*, pp. 293–299. UK: Elgar Reference Collection.

which are being updated continually. The latest updated version of guidelines was issued in 2006. These guidelines describe how biologics should be used in medical research and what type of precautions needs to be taken while using living beings as subjects in medical research. Guidelines talk about ethical concerns in human genetics, human genomic research, clinical trials of drugs, research in transplantation, assisted reproductive technologies, and things connected with such researches.[6] These guidelines are significant as biotechnology uses living beings as subjects in research and experimentation. In particular, application of biotechnology in the medical sector would attract the relevance of these guidelines. Based upon the continuous efforts of the ICMR in monitoring human genetic research and in formulating time to time updated guidelines "Assisted Reproductive Technologies Bill, 2010" was drafted and the same is pending before the Parliament.

Revised Guidelines for Research in Transgenic Plants and Guidelines for Toxicity and Allergenicity Evaluation of Transgenic Seeds, Plants and Plant Parts, 1998

These guidelines, issued by the Department of Biotechnology, Ministry of Science and Technology, Government of India in 1998, have been developed in the light of enormous progress that has been made in rDNA research and its widespread use in developing improved microbial strains, cell lines, and transgenic plants for commercial exploitation. The guidelines are considered to be a step ahead of DNA safety guidelines issued in 1990 (revised in 1994). They are meant for researches in plants involving rDNA, including the development of transgenic plants and their growth in soil for molecular and field evaluation. The issues regarding the import and shipment of genetically modified (GM) plants for research use are also covered. The institutional mechanism that was created for the regulation of biotechnology under the 1989 rules on microorganisms has been acknowledged and further empowered.[7] The

---

[6] 'ICMR Guidelines', 2006, Director General, Indian Council of Medical Research, New Delhi, available at http://www.icmr.nic.in/ethical_guidelines. pdf (last accessed 24 June 2016).

[7] See generally, Revised Guidelines for Research in Transgenic Plants, 1998.

various functions, scope, and need of such functions have been provided under the present revised guidelines on research in transgenic plants. The different categories of genetic engineering experiments on plants, as mentioned under the guidelines, could be discussed as follows.[8]

Category I

Under this category, routine rDNA experiments such as the following are covered: routine cloning of defined genes; DNA fragments of microbial, animal, and plant origin; non-coding stretches of DNA; transfer of defined cloned genes; research involving the study of transient expression in plant cells to study genetic transformation conditions; and open reading frames in defined genes, which were generally considered as safe to humans, animals, and plants. In the case of experiments, only intimation to the institutional biosafety committee is required.

Category II

This category includes the following lab and greenhouse or net-house experiments in a contained environment: defined DNA fragments that are non-pathogenic to human and animals used for genetic transformation of plants; experiments with marker genes; experiments with plants having resistance features to herbicides, biotic, and abiotic stresses; experiments with genes for the production of antibodies; experiments for gene tagging; and experiments with genes from different species. Experiments and research would need to be approved by Genetic Engineering Approval Committee.

Category III

This category pertains to high-risk experiments where the escape of transgenic traits into the open environment can cause significant alterations in the biosphere, ecosystem, plants, and animals by dispersing new genetic traits, the effects of which cannot be judged precisely. All experiments that are not falling under the other two categories

---

[8] See generally, Revised Guidelines for Research in Transgenic Plants, 1998.

would fall under this category. The experiments falling under this category can be conducted only after clearance from the Review Committee on Genetic Manipulation and notified by the Department of Biotechnology.

Guidelines for Stem Cell Issued By the ICMR in Association with Department of Biotechnology, Government of India, 2012

In March 2012, ICMR in association with Department of Biotechnology, Ministry of Science and Technology, Government of India issued "Guidelines for Stem Cell Research." These guidelines discuss range of issues pertinent to stem cell research in India, such as principles to be adhered to while embarking on stem cell research, minimum standards to be maintained, mechanism for review of research and responsibilities, various categories of stem cell research, clinical trials with respect to stem cell research, commercial exploitation, patenting, and international collaborations.[9] Biotechnology industry is growing in the country whereas there has been no regulatory means to monitor the research as such. It is believed that these guidelines would not only have the efficacy of monitoring stem cell research but it would regulate the legal aspects of stem cell research in India.

Guidelines and Standard Operating Procedures for the Conduct of Confined Field Trials of Transgenic Plants, 2008

The guidelines and the standard operating procedures (SOPs) for confined field trials of regulated, GE, and transgenic plants were issued by the Department of Biotechnology, Ministry of Science and Technology, Government of India in 2008. The guidelines provide for instructions to help applicants in meeting the requirements for authorization and approval of confined field trials of GE plants. These guidelines summarize the information, requirements, and procedures

---

[9] 'Stem Cell Research Guidelines', 2012. Director General, Indian Council of Medical Research, New Delhi, available at http://icmr.nic.in/stem_cell/stem_cell_guidelines.pdf (last accessed 24 June 2016).

used by the two regulatory committees, RCGM and GEAC. These two committees are responsible for evaluating and approving applications for confined field trials. The guidelines talk about contained and confined field trials for the purpose of clarity and appropriateness. The regulatory authorities for biotechnology in India, as provided under the 1989 rules, have been further recognized by the current guidelines. The Genetic Engineered Approval Committee, Review Committee on Genetic Manipulation, Institutional Bio-safety Committee, Recombinant DNA Advisory Committee, State Biotechnology Co-ordination/Committee, and the District-Level Committees, which constitute the institutional mechanism for the regulation of biotechnology in India, have been recognized and mentioned under these guidelines. Further, how the application seeking authorization to conduct field trials of GE plants be made is detailed. Besides, how such application would get through the Institutional Biosafety Committee to reach the Review Committee on Genetic Manipulation and the procedure involved therein have been specified. The general information for submitting application seeking authorization for conducting field trials, general requirements for confined field trials, prescribed restrictions, monitoring, mandatory submissions, and standard terms and conditions for authorization have been mentioned in detail. The SOPs have been prepared to provide guidance for conducting confined field trials of regulated and GE crops in India. These procedures provide guidelines for:[10]

- transport of regulated and GE plant material;
- storage of regulated and GE plant material;
- management of confined field trials;
- management of harvest or termination of confined field trials;
- post-harvest management of confined field trials.

A very detailed SOP has been provided under the guidelines regarding the aforementioned activities. The SOPs are considered to be very significant guidelines providing general requirements and activity-specific, detailed standards for confined field trials of GE plants.

---

[10] See generally, Guidelines and Standard Operating Procedures, 2008.

Guidelines for the Safety Assessment of GM Foods, 2008

To address the human health safety of foods derived from GE plants, there is a need to adopt a systematic and structured approach to their risk analysis. In the light of the international food regulatory commission Codex Alimentarius—the principles for the risk analysis of foods derived from modern biotechnology—the guidelines[11] aim for ensuring the safety of the GM foods in India. The guidelines provide for the assessment of possible toxicity and allergenicity of GM foods, and assessment strategy, screening, and analysis of key components of GM foods. The institutional mechanism for biotechnology regulation in India is recognized under the guidelines. The purpose of the Food Standards and Safety Act, 2006, and the various guidelines issued by the Department of Biotechnology, Government of India, are reiterated under these guidelines. With the increased use of GM foods, ICMR—the scientific and technical advisory body to the Ministry of Health and Family Welfare, Government of India—introduced these guidelines to establish safety-assessment procedures for GM foods in India, whereby international guidelines for the conduct of food-safety assessment derived from rDNA plants are taken into consideration. For the sake of clarity, GM plants, non-GM plants, genetic manipulation, characterization, and donor organisms have also been described. The assessment of toxicity, allergenicity, considerations for the safety assessment, assessment strategy, analysis of the key components, intended and unintended modifications, and effects resulted in the GM foods have also been discussed to provide significant assistance in assessing the safety of GM foods and their use in India.

Protocol for Safety Assessment of Genetically Engineered Plants/Crops, 2008

A series of protocols have been developed by the Department of Biotechnology as guidance to the applicants seeking approval for the environmental release of GE plants in India under the 1989 rules. They specifically address the key elements of the safety assessment of foods and/or livestock feeds that may be derived from GE crops.

[11] These guidelines have been issued by ICMR in 2008.

Institutional Mechanism for Biotechnology Regulation in India

Institutional mechanism is backed by three different legal norms as discussed above formulated by the Government of India, and is established to look into specialized issues pertinent to the regulation of biotechnology. The 1989 rules, as discussed earlier, define these competent authorities and their composition for handling the various aspects of the rules.[12] The DNA safety guidelines of 1990 also discuss authorities responsible for regulating various aspects of biotechnology. The revised guidelines of 1998 acknowledge the institutional mechanisms created for the regulation of biotechnology. Presently, there are six competent authorities in India. Let us discuss various authorities under the institutional mechanism in some detail.

Recombinant DNA Advisory Committee (RDAC)

The functions of RDAC are of an advisory nature, involving the review of developments in biotechnology at the national and international levels, and recommending suitable and appropriate safety regulations for recombinant research, use, and application from time to time. The committee functions under the Department of Biotechnology.[13]

Institutional Biosafety Committee

The Institutional Biosafety Committee (IBSC) is established under the institution or by the occupier engaged in GMO research to oversee such research and interface with the RCGM in regulating it. It is necessary that the organizations intending to carry out research activities involving genetic manipulation of microorganisms, plants, or animals should constitute their institutional biosafety committee.[14] The committee

---

[12] Official Gazette Notification No. GSIR 1037 (E) dated 5 December 1989 by the Ministry of Environment and Forests, Government of India.

[13] Section 4 of the 1989 rules.

[14] See generally, Revised Guidelines for Research in Transgenic Plants, 1998.

comprises the head of the institution, scientists engaged in DNA research, a medical expert, and a nominee of the Department of Biotechnology, Ministry of Science and Technology, Government of India. The occupier of the institution shall prepare an up-to-date on-site emergency plan according to the guidelines issued by RCGM and make its copies available to the state and district biotechnology coordinating committee and also to the genetic engineering approval committee. All recombinant research carried out by the institution should have a designated principal investigator whose duty would be to appraise the institutional biosafety committee about the nature of the experiments being carried out, and seek permission before starting the experiments.[15] If the risks associated with the research are considered to be of a higher magnitude having potential of polluting or endangering the environment, the biosphere, the ecosystem, the animals, and the human beings, permission from the RCGM would be required. The DNA safety guidelines state that in case of import and shipment of GE products and microorganisms, clearance from the institutional biosafety committee is required.[16] Furthermore, the Revised Guidelines for Research in Transgenic Plants, 1998, provide for further impetus to the institutional bioethics committee. According to these guidelines, institutional biosafety committee is the nodal point for interaction within an institute or university or commercial organization involved in rDNA research.

Review Committee on Genetic Manipulation (RCGM)

Established under the Department of Biotechnology, Government of India, the committee monitors the safety-related aspects of ongoing research projects, including small-scale field trials, and brings out the manuals and guidelines specifying the procedure for regulatory process with respect to activities involving GE organisms in research, use, and application, with a view to ensure environmental safety.[17] It lays down

---

[15] See generally, Revised Guidelines for Research in Transgenic Plants, 1998.

[16] Section 11 of the DNA Safety Guidelines, 1994.

[17] See generally, Revised Guidelines for Research in Transgenic Plants, 1998.

guidelines and protocols in this regard and also sanctions permission for experiments as well as the conditions under which the resultant product may be released. It also approves applications for generating research information on transgenic plants, and the guidelines on research in transgenic plants provide for further details in this regard. As per the guidelines on microorganisms read with DNA safety guidelines and also the revised guidelines on research in transgenic plants, the committee has the following functions:[18]

- It reviews the reports in all the approved ongoing research projects involving high-risk category and controlled field experiments.[19]
- It visits the sites of experimental facilities periodically where projects with biohazard potential are being pursued and also at a time prior to the commencement of the activity to ensure that adequate safety measures have been taken as per the guidelines.
- It issues clearance for import or export of etiologic agents and vectors, germplasms, organelle, and so forth, needed for experimental work or training and research.
- The committee is also entrusted with the task of bringing out manuals of guidelines specifying the procedure for regulatory process with respect to activities involving GE organisms in research, use, and applications, including industries, with a view to ensure environmental safety.
- All ongoing projects involving high-risk category and controlled field experiments are reviewed to ensure that adequate precautions and containment conditions are followed as per the guidelines.
- The procedures for restricting or prohibiting production, sale, importation, and use of such GE organisms are laid down by the committee.[20]

---

[18] See generally, DNA Safety Guidelines, 1994.

[19] See the 1989 rules.

[20] The Review Committee on Genetic Manipulation shall include representatives of the Department of Biotechnology, Indian Council of Medical Research, Indian Council of Agricultural Research, and Council of Scientific and Industrial Research.

Genetic Engineering Approval Committee

Established under the Ministry of Environment, Forest, and Climate Change, Genetic Engineering Approval Committee (GEAC) is the apex body to accord approval of activities involving large-scale use of hazardous microorganisms and recombinants in research and industrial production from the environmental angle. It is responsible for granting approvals related to the release of GE organisms and products into the environment, including experimental field trials.[21] Further, it approves the import, export, transport, manufacture, process, use, and selling of hazardous microorganisms or GE organisms or substances or cells.[22] It gives directions to the occupier and institution to take measures concerning the release of microorganisms or GE organisms or cells from the laboratories, hospitals, or such other areas. It can also prohibit and control such release by laying down measures to be followed. Any person operating or using GE organisms or microorganisms will have to obtain license from the committee for any such activity. However, for educational purpose, certain experiments are allowed within the field of gene technology or microorganisms outside the notified laboratories[23] but under the supervision of institutional biosafety committee. Any large-scale industrial processes and operations should be taken up only after the approval of the committee.[24] Furthermore, any production of GE organisms or cells or microorganisms should not be commenced without the consent of the committee.[25] Similarly, food items, ingredients in food items, and additives including processing and containing or consisting of GE organisms or cells shall not be produced, sold, imported, or used except with the approval of the committee.[26] Before granting such an approval, the committee seeks information from the people working on microorganisms, examines, and inspects on-site

[21] Section 7 of DNA Safety Guidelines, 1989.

[22] Section 7 of the 1989 rules.

[23] It is to be noted that in general, experiments concerning microorganisms or GE organisms are supposed be conducted only in the notified laboratories under the Environment (Protection) Act, 1986.

[24] Section 7 of the DNA Safety Guidelines, 1994.

[25] Section 8 of the 1989 rules.

[26] Section 11 of the 1989 rules.

emergency plans in the laboratories or other such places. Any change in information shall be notified to the committee without delay.[27] While granting the approval, the committee can prescribe the terms and conditions to the applicant and maintain control and supervision over the use of microorganisms. Implementation of such terms and conditions can be supervised by the committee through state biotechnology coordination committee or state pollution control boards or the district-level committees.[28] Such approvals are generally granted for a period of four years, subject to renewal.[29] Once granted, the approval can be revoked on the evidence of adverse conditions that are harmful to the environment, nature, or health. It is clearly mentioned under the rules that deliberate or intentional release of GE organisms or hazardous microorganisms, or cells, including deliberate release for the purpose of experiment, is not allowed,[30] unless approved by the committee.[31] The various functions of the committee, as discussed earlier, have also been endorsed under the DNA Safety Guidelines, 1994.[32] Regarding the post-harvest handling of transgenic plants, the guidelines state that the committee will have to monitor the process of destroying and burning all the vegetative parts and left-over seeds.[33] In case of release of microorganisms and GE products into the environment, the guidelines provide for the process of risk assessment for enabling the committee to prescribe the terms and conditions to ensure safety and security of the environment and health. The guidelines state that a number of factors—geographical location, nature of release, on-site safety procedures, genetic stability, growth and survival characteristics of the host organism, ecosystem, nature of genetic modification, and so forth—should be considered by the committee for risk assessment.[34] The guidelines further provide for quality

---

[27] Section 12 of the 1989 rules.

[28] Section 14 of the 1989 rules.

[29] Section 13 of the 1989 rules.

[30] Deliberate release shall mean any intentional transfer of GMOs or hazardous microorganisms or cells to the environment or nature, irrespective of the way in which it is done.

[31] Section 9 of the 1989 rules.

[32] See generally, DNA Safety Guidelines, 1994.

[33] Section 9 of the DNA Safety Guidelines, 1994.

[34] Section 10 of the DNA Safety Guidelines, 1994.

control of biologicals produced by DNA technology. In this regard, it is required that the scientific details of methodology used for genetic engineering, cloning, construction of vectors, host type, origin of host DNA, or nucleotide sequence, should be provided to the committee.[35]

State Biotechnology Coordination Committees (SBCCs)

This committee has a major role in monitoring the use and exploitation of biotechnology. As prescribed under the DNA Safety Guidelines, it also has the power to inspect, investigate, and take punitive action in case of violations of statutory provisions through the nodal department and the state pollution control board or the Directorate of Health or Medical Services.[36] Periodically, it also reviews the safety and control measures in the various industries or institutions handling GE organisms or hazardous microorganisms.[37] The rules state that the use of microorganisms or GE organisms or cells for the purpose of research is allowed in laboratories, as notified by the Ministry of Environment and Forests under the Environment Protection Act.

District-Level Committees (DLCs)

The district-level committees have a major role in monitoring the safety regulations in institutions engaged in the use of GMOs/hazardous microorganisms and their applications in the environment.[38] As per the DNA Safety Guidelines[39] issued by the Department of Biotechnology, Government of India, the committee can visit these installations, formulate information chart, find out the hazards and risks associated with each of these installations, and coordinate activities with a view to meeting any emergency. The committee regularly submits reports to the state biotechnology coordination committee and also to the genetic engineering approval committee.

[35] Section 12 of the DNA Safety Guidelines, 1994.
[36] Section 4 of the 1989 rules.
[37] See generally, DNA Safety Guidelines, 1994.
[38] Section 4 of the 1989 rules.
[39] See generally, DNA Safety Guidelines, 1994.

Guidelines for Generating Preclinical and Clinical Data for rDNA Vaccines, Diagnostics, and Other Biologicals, 1999

The Department of Biotechnology, under the Ministry of Science and Technology of the Government of India, issued these guidelines in 1999. Given the fact that the products developed through rDNA technology are available in the market and more such products would be available in the market in the near future, there is a need to set regulatory standards for preclinical evaluations and assessment of such products. There are prescribed standards for conducting preclinical evaluation and assessment under the guidelines, whereby the therapeutic efficacy of vaccines and the toxicology of the derived products are assessed and the specification and characterization of information on rDNA vaccines and biological products are done. On the whole, the guidelines aim to ensure the safety, purity, potency, and efficacy of rDNA-derived products and to generate clinical data for rDNA vaccines, diagnostics, and other biologicals. The data generated is submitted to the Drug Controller General of India, who can use it while considering the release of drugs, vaccines, pharmaceuticals, and other such diagnostic products into the market.

Indian Council for Medical Research Guidelines on GMO, 2007

The ICMR came up with its draft guidelines pertaining to Safety Assessment of Foods Derived from GE Plants in 2007. The ICMR, in its capacity as the scientific and technical advisory body to Ministry of Health and Family Welfare, Government of India, has formulated these guidelines to establish the safety-assessment procedures for foods derived from GE plants, taking into consideration the international Guideline for the Conduct of Food Safety Assessment of Foods Derived from RDNA Plants.[40] While the objective is to determine whether

---

[40] It means that the international guidelines as mentioned in the text are formulated by Codex Alimentarius Commission (CEC) in the year 2003. Codex guidelines have been taken into account while drafting ICMR food guidelines. Available at http://icmr.nic.in/final/rda-2010.pdf (last accessed 24 June 2016).

the GM food presents any new or greater risks in comparison with its traditional counterpart, or whether it can be used interchangeably with its traditional counterpart without affecting the health or nutritional status of consumers, the inherent objective is to establish the relative safety of the new product such that there is a reasonable certainty that no harm will result from intended uses under the anticipated conditions of processing and consumption.[41]

Protection of Plant Varieties and Farmers' Rights Act, 2001

The act basically talks about promotion and protection of new and innovative varieties of plants and protection of farmers' rights in India. At the same time, the act also talks about prohibition of controversial genetic research on plants. For instance, the act prohibits genetic research developing terminator technology in plants.[42] Terminator technology is developed through genetic research on plants to make plants produce sterile seeds and those plants are useful only for human consumption. Therefore, Teminator technology would put off intended plants' natural trait to produce seeds which can be further sowed.[43] Such type of controversial technologies developed through biotechnology is prohibited from commercial exploitation and private monopoly under the legislation.

---

[41] "If a new or altered hazard, nutritional or other food safety concern is identified by the safety assessment, it is further evaluated to determine its relevance to human health. Following the safety assessment and, if necessary, further risk analysis, the food or component of food may be subjected to risk management options before it is considered for commercial distribution. Where no conventional counterpart exists for comparison, the safety of a GM food must be evaluated from data derived directly from historical experience or experimental studies with the food." Available at http://www.icmr.nic.in/guidelines.htm (last accessed 3 March 2016).

[42] Sreenivasulu, N.S. 2011. *Intellectual Property Rights*, p. 63. New Delhi: Regal Publications.

[43] Sreenivasulu, N.S. 2013. *Law Relating to Intellectual Property*, 1st edn. Bloomington, Indiana, USA: Penguin-Partridge Publications.

Patent (as amended in 2005) Act

Patent Act of India, 1970, prohibits private monopoly on living beings such as microorganisms, plants, and animals, keeping them outside the purview of patent subject matter.[44] However, through amendments made in the Patent Act in 1999, 2002, and 2005, certain biotechnology innovations, including GM living beings such as microorganisms, have been allowed for patenting.[45] It has been a milestone amendment for biotechnological field.

The Biodiversity Act, 2002

Biological resources are the raw material for biotechnology research and experimentation. It goes without saying that naturally available living resources are changed to produce nonnatural and engineered living beings such as GE microorganism, plant, animal, and human genetic material.[46] In this context, the Biological Diversity Act talks about prior consent to be taken from the government before using such natural biological resources for genetic engineering and such other experimentation.[47] The act also talks about sovereign rights over biological resources for the host country and also about sharing of benefits in case of successful commercial exploitation of innovation produced through biotechnology while using biological resources. Therefore, mandates given in the act need to be followed at the level of experimentation and also at the stage of commercial exploitation of biotechnology innovation.

Guidelines for Evaluation of Probiotics in Food, 2011

These guidelines have been formulated by the Department of Biotechnology, Government of India, and the ICMR. In the

[44]   Sreenivasulu 2013. *Law Relating to Intellectual Property*, pp. 215–18.

[45]   Sreenivasulu. 2011. *Intellectual Property Rights*, p. 30.

[46]   Sreenivasulu, N.S., K.S. Kariyanna, and B.S. Viswanath. 2012. 'Biological Diversity, Intellectual Property and Patents: Concerns of Biological Resources', *Manupatra Intellectual Property Reports*, 1(2): F47–F58.

[47]   Sreenivasulu, N.S. and Arnab Sengupta. 2010. 'Biological Resources, IPR and Biodiveristy', *Manupatra Intellectual Property Reports*, 1(4): F35–F46.

contemporary times, the use of probiotics in food has increased compared to a situation a decade or two earlier. The global probiotics market was worth USD 15.9 billion in 2008.[48] By 2014, it was expected and projected to double from USD 15.9 billion to USD 32.6 billion. Similarly, in India, the probiotic industry is estimated to be worth USD 20.6 million by the end of 2015.[49] In India, there are no regulatory guidelines for probiotic foods. In the absence of any such standards and guidelines, there is a great scope for spurious products with false claims being marketed. Therefore, it was mandatory that probiotic products fulfil some essential prerequisite conditions before being labeled as a "probiotic product." A holistic approach was therefore needed for formulating guidelines and regulations for evaluating the safety and efficacy of probiotics in India, which should be in consonance with current international standards. Thus, the guidelines have been developed with the aim to guide the regulatory authority for evaluating probiotic products in India. The international standards on food safety and standards, as prescribed by Codex Alimentarius Commission, are quite relevant here. The commission mandates for food hygiene and provides for the guidelines for application of hazard analysis and critical control point.[50] Safety measures and guidelines on food hygiene prescribed by the Codex Commission have been referred to under the current guidelines for necessary follow-up. The labeling requirement, as prescribed by the international and national legal agencies, has been made stricter and stringent for the use of probiotics in food. On the whole, the guidelines deal with the use of probiotics in food and provide the requirements for safety, assessment of safety

---

[48] Frost and Sullivan Research Service. 2009. 'Probiotics in Food and Beverages: A Strategic Assessment of Indian Market,' available at http://www.frost.com/prod/servlet/report-brochure.pag?id=P35E-01-00-00-00 (last accessed 2 March 2016).

[49] Indian Council of Medical Research. 2011. 'ICMR-DBT Guidelines on the Evaluation of Probiotics in Food,' available at http://icmr.nic.in/guide/PROBIOTICS_GUIDELINES.pdf (last accessed 2 March 2016).

[50] Reid, G. 2001. 'Regulatory and Clinical Aspects of Dairy Probiotics,' Background paper for FAO/WHO Expert Consultation on Evaluation of Health and Nutritional Properties of Probiotics in Food including Powder Milk and Live Lactic Acid Bacteria, Cordoba, Argentina, 1–4 October.

and efficacy of the probiotic strains, and health claims and labeling of products with probiotics.

DNA Profiling Bill, 2015

DNA profiling and analysis of body substances is considered to be a powerful technology that makes it possible to determine whether the origin of source of one body substance is identical to another. It is also helpful in establishing the biological relationship between individuals, living or dead. The Indian government has been attempting to introduce regulations on DNA profiling for a decade now. The first culminated effort in this regard was made in 2007 when DNA profiling bill of 2007 was prepared with the help of the Department of Biotechnology, under the Ministry of Science and Technology, Government of India. When the bill was made public, many non-governmental organizations (NGOs) and activists raised their voice against the bill. Hence, the 2007 version of the bill was never introduced in the parliament. Instead, it was sent to the Department of Biotechnology and the Center for DNA Fingerprinting and Diagnostics (CDFD), Hyderabad, for updation. In 2010, the new draft of DNA profiling bill was prepared and further updated in 2012 and again in 2015.[51] The current draft of the DNA profiling bill of 2015 is further updated and proposes to have a national DNA databank and an authority to regulate the collection, maintenance, and use of DNA profiles in India. It also states the purpose for which DNA samples and profiles can be used and intentions behind the collection of DNA profiles and samples. The act seems to be in the lines of US legislation, the Genetic Information Nondiscrimination Act.

What is DNA Profiling?

It is an analysis of body substances at the DNA level to determine whether the source of identification of one body substance is identical to another, and establishes the relationship between two individuals,

---

[51] Initially in 2007, DNA profiling bill was proposed that never became an act. Later in 2012, another draft of DNA profiling Bill was made and in 2015 once again another draft was prepared which is again awaiting the parliament's approval to become an enactment.

living or dead, without any doubt. It involves collection of a few skin cells, a muscle tissue, a hair root, blood, saliva samples, and even the objects used by the individuals such as a spoon, toothbrush, clothes, or glass. From these samples collected by authorized professionals,[52] DNA strands are extracted to establish genetic identity.

DNA Profiling Board

It is proposed under the bill that DNA Profiling Board would regulate the issues pertinent to profiling of DNA, its maintenance in the data banks, its use, and indexing. The board consists of members[53] of high integrity and moral, who are nominated by the President, including a population geneticist. The board functions from its headquarters in Hyderabad, India. It is proposed to be headed by a biologist and members from the fields of law and police personnel. The bill states that the board has the power to specify the process, sources, and manner of DNA sample collection and profiling, and can be approached by any aggrieved person covered under the bill. If the board fails to take any action or when the board action is not acceptable to the aggrieved party, the central government can be approached.[54]

DNA Laboratory

For storing, profiling, and analyzing the DNA sample, there is a requirement to establish DNA laboratories. The infrastructure, staff, and maintenance of laboratories would be regulated by the DNA Profiling Board. The approval for the establishment, certification, and functioning of the DNA laboratories come from the board. The standards, quality controls, and duties of these laboratories are fixed and shall be regulated by the board.

National DNA Databank

Profiling the DNA samples of the people would help in the national interest in terms of helping the prosecution to identify the suspected.

---

[52] Section 53 and 54 of the Criminal Procedure Code.
[53] Section 4 of the Criminal Procedure Code.
[54] Section 18 of the DNA Profiling Bill, 2015.

This can happen by collecting the DNA samples from the crime areas and matching the same with the DNA databank[55] of the people. Similarly, if the suspected is from any other nation, the collected samples of DNA can be matched with the DNA databank of that country with the help of that country's prosecution. It would be possible only if countries maintain national databanks of DNA profiles through an established DNA profiling regulatory authority. The state labs would do the DNA profiling and create genetic data that would be transferred and maintained with the national DNA bank. The databank manager can grant permission to any individual or government agency to have access to the DNA profile in appropriate cases. The access to the DNA database can be restricted in the public interest.[56] Furthermore, DNA databank manager can be called upon by any court or tribunal for testifying the DNA data, DNA profiles and its availability in the existing data with the bank, its suitability, and relevance for the issue at hand.[57] Whenever it is required, DNA profiles maintained with the national DNA bank can also be shared with the foreign countries.[58] However, confidentiality, privacy, and secrecy of the DNA profiles and data shall be maintained by the bank, whereby the profiles and data maintained shall only be used for the purpose of identifying persons in criminal and civil matters.

Purpose for Which DNA Profiles Can be Used

DNA profiling is useful in identifying the culprits in criminal offenses, confirming how and whether people are related to each other, paternity testing, and identifying dead bodies and missing persons. Scientifically speaking, human beings get 50 percent of their DNA from the mother and the remaining 50 percent from the father. Such being the case, DNA profiling would help in identifying the parents in cases of deserted children or orphans, and would be useful in disposing of paternity suits. In a case of DNA profiling for identifying suspects in a criminal offence, one can also demand DNA testing to prove their innocence before the court of law. The purposes for which DNA profiling can be used is provided under the

---

[55] Section 24 of the DNA Profiling Bill, 2015.
[56] Section 36 of the DNA Profiling Bill, 2015.
[57] Section 26 of the DNA Profiling Bill, 2015.
[58] Section 29 of the DNA Profiling Bill, 2015.

schedule to the act. The board has the power to specify instances and purposes when and which DNA samples and profiling are to be used.[59] The central government has the power over and above the board to specify instances when DNA samples and profiling can be used.[60]

Instances Where DNA Profiling is Required

In cases of offences against human body, DNA profiling is required to deliver justice. It is stated[61] that for investigating offences under the Indian Penal Code, Protection of Civil Rights Act, and other such legislations that empower the state for criminal prosecution, DNA profiling could be done. The instances that necessitate DNA profiling have also been specified.[62] In legal cases, DNA profiling can be done compulsorily even without the consent of the individual. In other cases, the consent of the individual is required. At the same time, it is mentioned that in cases of civil disputes also, DNA profiling can be done as and when it is necessary. It is categorically stated that for establishing individual's identity and issues relating to immigration, it is possible to do justice through DNA profiling. Furthermore, the board can decide which civil matters require DNA profiling. The bill aims to modernize crime detection, conviction, and acquittal; in total, it would revolutionize crime investigation, litigation, and adjudication. The Schedule to the DNA bill provides a list of applicable instances/matters of human DNA profiling.[63] In a number of offences under Indian Penal Code, Hindu Marriage Act, Domestic Violence Act, Dowry Prohibition Act, and Motor Vehicle Regulation Act, DNA profiling is done to assist the judicial adjudication process to arrive at a proper conclusion. Thus, DNA profiling is required in case of the following offences:

- culpable homicide[64]
- murder

---

[59] Section 13 and 58 of the DNA Profiling Bill, 2015.
[60] Section 57 of the DNA Profiling Bill, 2015.
[61] Schedule to the DNA Profiling Bill, 2015.
[62] Schedule to the DNA Profiling Bill, 2015.
[63] Shedule to the DNA Profiling Bill, 2015.
[64] Under the Indian Penal Code.

- death caused by negligence
- dowry death[65]
- hurt and grievous hurt
- wrongful restraint
- wrongful confinement
- criminal force or assault
- kidnapping
- abduction
- rape[66]
- sexual assault
- prevention of child's birth or feticide
- death of new-born child or infanticide
- unnatural offences such as sodomy and bestiality
- outraging the modesty of women
- accidents[67]
- cohabitation with women by deceit
- adultery
- eliciting married women with criminal intent
- cruelty to married women[68]

Sources of Collection for DNA Test

The sources of collection of DNA test have been specified under the bill. The schedule to the bill lists the sources of collection of samples for DNA test. Section 2 read with Section 66 of the bill provide a list of instances where samples could be collected for DNA test.[69] The following instances can be the sources for the samples:

- scene of occurrence or crime
- medical examination
- autopsy examination
- exhumation

[65] Under Dowry Prohibition Act.
[66] Under the Indian Penal Code.
[67] Under Motor Vehicles Act.
[68] By husband or wife or family members under Domestic Violence Act.
[69] Section 2(1) and Section 66(zh) of the DNA profiling Bill.

- tissue and skeleton remains
- clothing and other objects
- preserved body fluids and other samples
- intimate body samples[70]
- non-intimate body samples[71]

The Other Side of the Coin

It is observed by scholars that DNA profiling may not be completely foolproof. Further, several studies supported by investigation database have argued that there are cases where DNA profiling has failed to detect suspects of a crime; on the contrary, this procedure may be used for illegal purposes such as identification of the medical history, family history, caste, even the actual location of an individual.

DNA for Data Storage

The evolution of storage media has seen a wide range of devices in the past few decades—from floppy disks with 80-kilobyte storage capacity to the present-day portable drives that holds terabytes of data. It might be surprising to know, however, that today, the evolution of data-management techniques is powered by DNA for storing data. Just one gram of DNA is capable of holding 455 exabytes, which is more than enough to store for all the data held by Google, Facebook, and every other major technology companies.[72] Seen as a durable mode of storage, DNA is extracted and sequenced from 700,000-year-old horse bones.

---

[70] Intimate body samples from living persons shall be collected and forensic procedure shall be performed by a registered medical practitioner, as defined under Section 53 of the Criminal Procedure Code. The medical practitioner familiar with DNA sample collection procedure required to be followed by the laboratory to which the samples have to be forwarded.

[71] Non-intimate body samples can be collected and non-intimate forensic procedure can be followed by the technical staff trained for the collection of samples for DNA test under the supervision of a medical officer or scientist or by other person as specified in the regulation.

[72] *The Times of India*. 2015. 'DNA Drives Data Storage,' Kolkata, 23 February.

Swiss scientists experimenting with DNA storage have found that the data in DNA can last for 2,000 years if kept at a temperature of around 10 degree Celsius.[73] These experiments and their possible results can bring revolutionary changes in data storage and management.

Biotechnology Regulatory Authority: Relevance and Need for National Regulation

The Biotechnology Regulatory Authority Bill, 2008, has been pending in the parliament for several years now. In fact, though the bill was promising, it lapsed in the parliament because it could not get the consensus among the legislators in both the houses of the parliament. Further, in 2013, another draft of the bill—Biotechnology Regulatory Authority Bill, 2013—was introduced. As India is promoting biotechnology and its use for different purposes, it is felt that a single body is needed that would regulate different issues pertinent to the promotion, use, and exploitation of biotechnology. As experienced, biotechnology can be an answer for ever-increasing food needs of the country. It has been recognized that without biotechnology and GM crops, it is not possible to ensure food security and it is being reflected in the National Food Security Act of 2013 also. Further, for medical and therapeutic purposes, biotechnology is very useful. Likewise, biotechnology can be commercially used and promoted in various industries. Having recognized the significance and varied applications of biotechnology, it was felt necessary to have the National Biotechnology Regulatory Authority through the National Biotechnology Regulatory Authority of India Bill, 2013. It is felt that with the Government of India being close to allowing field trials for GM crops, advancement in the field of biotechnology products can be expected. Recently, the Genetic Engineering Approval Committee functioning under the Ministry of Environment and Forests has cleared field trial approvals for few GM crops including rice and brinjal. However, there are some cultural and social objections and concerns for the same in India. For this reason, in 2012, the government imposed moratorium on conducting field trials of GM crops in India including Bt rice and Bt brinjal. Amidst doubts whether India is ready for GM food or not, the debate on whether

---

[73] For more details, refer to newsscientist.com (last accessed 3 March 2016).

to promote Bt or not to promote Bt is still intense, with the Government of India lifting an 18-month freeze and clearing field trials of at least 13 GM food crops, including the contentious mustard and brinjal.[74] The issue of promoting biotechnology regulation and GM crops in India is a much debated, and has now become a sensitive and controversial matter. Keeping in mind the enormous social and economic influence of biotechnology, a consolidated and codified set of legal framework needs to be set up to regulate various aspects of biotechnology. This can be done through the pending Biotechnology Regulatory Authority of India Bill after arriving at a consensus with the various stakeholders of biotechnology industry and application. If not, there shall be a fresh attempt to put up an acceptable legal framework for regulating biotechnology in India.

Parliamentary Standing Committee on Agriculture of the Government of India

Given the enormous challenges that farmers and plant breeders face in dealing with the conservation and sustainable management of essential crop germplasm, in order to address the issues of food security and malnutrition, particularly among some of the world's poorest nations, it is hardly surprising to find GE or Bt crops being grown in countries such as Argentina, South Africa, China, Philippines, and India.[75] Biotechnology is believed to possess the potency to increase productivity and enhance nutritions in food crops. It is also believed that biotechnology is capable of catering to the food security needs of the masses.[76] However, this potency of biotechnology in enhancing nutritional values in food is not away from extensively debated controversies and fears.

---

[74] Sahai. 2014. 'We do not have the competence to play around with GM foods,' *The Times of India*, Kolkata, 2 November, available at http://timesofindia.indiatimes.com/home/sunday-times/all-that-matters/We-do-not-have-the-competence-to-play-around-with-GM-foods-Suman-Sahai/articleshow/45009714.cms (last accessed 25 June 2016).

[75] Francioni, Francesco and Tullio Scovazzi (eds). 2006. *Biotechnology and International Law*, p. 258. Portland, USA: Hard Publishing.

[76] Sreenivasulu, N.S. and Debanshu Chettery. 2016. 'Agricultural Biotechnology: Environmental Law and Other Policy Considerations,' *Environmental Law Practice Review*, (I): 25–39.

Nevertheless, having recognized the importance of resolving the issue of whether to promote biotechnology in full steam by allowing GM crops for field trials, cultivation, and human consumption, the Indian parliament formulated a standing committee on agriculture in 2009 to look into, advice, and submit a report in this regard. The Ministry of Science and Technology,[77] in their submission to the committee, stated that ever since domestication of crop plants 10 millennia ago, man has endeavored to improve the productivity of crop plants. Advances in mineral nutrition, irrigation, and plant breeding have led to a series of revolutions resulting in increased crop productivity and ensuring food security to the ever-increasing population worldwide. India's population is expected to reach 1.5 billion by 2025, making food security the most important social issue; further, food production will have to be increased considerably to meet the needs of the growing population. There are a number of terms that are referred to the committee by the Government of India through the Ministry of Agriculture and Farmers Welfare and the Ministry of Science and Technology. These include:

- growing population of the countries;
- persistent problems in the agricultural sectors;
- innovations in the agricultural biotechnology;
- success of biotechnology in the developed nations and global scenario;
- adaptability of biotechnology and GM crops in India;
- prospects and effects of GM crops in India.

The parliamentary standing committee on the cultivation of GM crops has suggested the stopping of all the field trials on GM crops and complete overhaul of regulatory norms. The standing committee on agriculture deliberated for 3 years—from 2009 to 2012—studying and considering the international norms and regional and domestic laws. The stakeholders from different fields such as agriculture, biotechnology industry, political parties, nongovernmental and people's organizations, social activities, scientific community, academicians, lawyers, and other

---

[77] Available at www.dst.gov for latest updates (last accessed 3 March 2016).

interested people and agencies were consulted in preparing the report[78] and submitted the same to the Ministry of Agriculture and Farmers Welfare on 7 August 2012. Moratorium on GM crops, in particular Bt brinjal in 2012, took place in the background of the parliamentary committee's findings that India was not yet prepared and ready for GM crops.[79] It had been brought to the notice of the committee that India was the second largest producer of brinjal in the world after China, which sparked debate whether India really needed Bt variety of brinjal at all when there was substantial amount of production of brinjal in the country.

Indian Judiciary on GM Crops

In *Aruna Rodrigues and Ors v. Union of India*,[80] the petition was filed opposing the approval of field trials for GM crops by the government. The petitioner sought Supreme Court's intervention in the matter with a clear direction to the government for not pushing and going ahead with the approval of field trials of GM crops. The petitioner's argument was as follows:

> The fact is that we have little that can be called rigour or comprehensive regulation on biotechnology and GM crops and their field trials. We don't even have the expertise in the areas needed. The scathing indictment of the regulators with regard to their oversight of the very first stage of risk assessment, i.e., the molecular analyses is clearly present. It is disquieting indeed that the government intends to introduce GM crops with the full realization that there will be no segregation of GM produce and no labeling for GM content. Government is attempting to railroad GM crops into India in a manner which is unconstitutional, contrary to our federal structure, contrary to the science, and which is suggestive of an hidden agenda, with suspicions of cronyism and corruption of a kind for which Monsanto in particular has become notorious worldwide.

---

[78] Parliamentary Standing Committee report was submitted to the Ministry of Agriculture and Farmers Welfare on 7 August and placed before both the houses of Parliament on 9 August 2012.

[79] See also views expressed by Suman Sahai. 2016. 'We do not have the competency to play around with GM foods,' p. 16.

[80] Writ Petition (Civil) No. 260 of 2005, *Aruna Rodrigues v. Union of India*.

The petitioner further states:[81]

> Any bio-security/GM-triggered failure will devastate our agriculture and
> export markets which are growing exponentially, including a Rs 12,000
> crore export market for rice, and agro-ecological farming systems. This is
> also in the light of the firm evidence that higher yield and reduced pes-
> ticide use in GM crops is a myth; it is conventional breeding methods
> that are delivering real increases in yield. We must look to investing in
> modernizing agro-ecological farming methods for food security.

While deliberating on the issue of field trials of GM crops in India,
the Supreme Court of India thought that there was a need for deeper,
wider, and broader view to be taken on the issue, supported by scientific, technical, and social elements. For this purpose, a Technical Expert
Committee (TEC) was constituted by the Supreme Court in July 2012.
Along with issues concerning biosafety, the extent of the contribution
of biotechnology and genetic engineering to the pressing issues of
food security and the right to adequate and nutritious food had to be
established.[82] In this background, the committee of technical experts,
comprising scientists from top public research laboratories and academic
institutions, have changed the 10-year moratorium on field trials of Bt
transgenics that it recommended in October 2012 to what appears to
be an indefinite moratorium on food crops in its final report. In fact,
the committee recommends for indefinite moratorium on Bt crops till
independent review on herbicide-resistant crops and crops for which
India is the center of origin/diversity. The committee suggested for
radical regulatory reforms and the examination/study of the safety dossiers. It was observed by the committee that there were major gaps in
the regulatory system.[83] These gaps needed to be addressed before the
issues related to tests could be meaningfully considered. "Till such time,
it would not be advisable to conduct more field trials," the experts said in
their final report without specifying any time frame. In other significant

---

[81] Writ Petition No. 260 of 2005, *Aruna Rodrigues v. Union of India.*
[82] Francioni and Scovazzi. 2006. *Biotechnology and International Law*, p. 258.
[83] Available at http://www.business-standard.com/article/economy-policy/
gm-crops-pm-revealed-his-assertive-self-to-push-for-trials-114041400028_1.
html (last accessed 9 August 2014).

recommendations, the panel found that herbicide-tolerant (HT) crops were "completely unsuitable"[84] in the Indian context and recommended that field trials and release of HT crops should not be allowed in India. Noting that single committees in the form of the main regulatory agencies for biotechnological crops—GEAC and RCGM—doing all the evaluation is not sufficient, the expert panel called for the setting up of a secretariat comprising dedicated scientists with area expertise as well as expertise in biosafety: "[t]his will require consultation with experts having experience at the international level in biosafety testing and evaluation of GM safety dossiers in reputed regulatory bodies;" it suggested that this should be done in collaboration with the Norwegian government. The reason for choosing Norway was that the Norwegian system had "an established commitment"[85] and was one of the few attuned to considering socioeconomic issues that would be important in the Indian context. The report said that the new regulatory body should have area-wise subcommittees/expert groups in the following fields: health (human and animal), environment and ecology, agro-economics and socioeconomics, molecular biology, soil science and microbiology, plant biology, and regulatory toxicology, among other specializations. Since the single largest number of applications for field trials to GEAC was for Bt transgenics, including food crops such as rice, the scientists were of the view that the safety of Bt transgenics with regard to chronic toxicity needed to be established before they could be considered safe for human consumption. In this regard, it pointed out that the largest deployment of transgenics worldwide was in soybean, corn, cotton, and canola, all of which were used primarily for oil or feed after processing: "[n]owhere are BT transgenics being widely consumed in large amounts for any major food crop that is directly used for human consumption."[86]

---

[84] Jisnu, Latha. 2013. 'Indefinate moratorium on GM field trials recommended', 22 July, available at http://www.downtoearth.org.in/news/indefinite-moratorium-on-gm-field-trials-recommended-41730 (last accessed 25 June 2016).

[85] Jisnu. 2013. 'Indefinate moratorium on GM field trials recommended'.

[86] *The Economic Times.* 2013. 'Put genetically-modified crop trials on hold for now: Supreme Court Panel', 23 July, available at http://articles.economictimes.indiatimes.com/2013-07-23/news/40749288_1_open-field-trials-bt-food-crops-interim-report (last accessed 25 June 2016).

The committee recommended that there should be a moratorium on field trials for Bt in food until there was more definitive information from sufficient number of studies as to the long-term safety of Bt in food crops. Perhaps, the GEAC had approved the commercial release of Mahyco's Bt brinjal in 2009, but the then environment minister Jairam Ramesh had put a moratorium in 2011–12 on its release in the wake of widespread public protests against the first transgenic food crop in the country. If TEC's recommendations were accepted, crops that originate in India, such as brinjal, could not be genetically modified. According to the panel: "[t]o date, no GMO that is intended primarily and directly for food production has been commercially released into its centre of origin."[87] It notes that the US has restrictions on the growth of Bt cotton in Hawaii where a weed related to cotton is found. For good measure, it emphasizes that cotton is not even a food crop. Crops in their centers of origin and diversity often have "a deep cultural significance that can easily get lost when utilitatarian issues dominate the discourse," says the 94-page report.[88] Ceremonial and medicinal varieties can also be put at risk from GM crops by reduction of diversity and genetic purity, and to justify their release, "there needs to be extraordinarily compelling reasons and only when other choices are not available. GM crops that offer incremental advantages or solutions to specific and limited problems are not sufficient reasons to justify such release."[89] In the present circumstances, there is no such compulsion, according to the scientists, who were categorical that the release of GM crops for which India is a center of origin or diversity should not be allowed.[90] Perhaps, the committee has conducted a detailed probe into the issue that was referred to by the Supreme Court of India. Taking note of every possible precursor, post-effect, and implication of field trials of GM crops,[91] the committee has suggested improvizing the regulatory structure and the governing

[87] Writ Petition No. 260 of 2005, *Aruna Rodrigues v. Union of India*.

[88] Jisnu. 2013.'Indefinate moratorium on GM field trials recommended'.

[89] Jisnu. 2013.'Indefinate moratorium on GM field trials recommended'.

[90] Available at http://www.gmwatch.org/index.php/news/archive/2013/ 14866-indefinite-moratorium-on-gm-field-trials-recommended-in-india (last accessed 9 August 2014).

[91] See, generally, http://www.gmwatch.org/index.php/news (last accessed 10 August 2014).

mechanism for the purpose of ensuring safety, security, and health before allowing field trials and encouraging commercial cultivation. It was felt that there was a need to prepare a perfect platform before starting the show and its showcasing. Without proper preparation and a platform for facilitating safe and secure field trials and commercial cultivation, there could be alarming risks and dangers associated with such unassessed and deliberate approval of field trials of GM crops. In the light of the discussed report of the technical committee, the petitioner has submitted the following report before the Supreme Court:[92]

> It is significant that this report of the technical experts committee, cautioning the court on the absolute need for rigorous safety regulation, comes at a time when in the last 72 hours, Monsanto has withdrawn applications for GMOs in the EU[93] because of non-acceptance on scientific grounds and rejection by civil society. It is also noteworthy, that the technical experts' committee report is the 4th official report which finds common ground on the lack of integrity and independence in the regulator and the presence of a pervasive conflict of interest which make sound and rigorous regulation of GMOs impossible. It is the 3rd official report barring GM crops/field trials singly or collectively. Furthermore, it is now no longer necessary to prove that the GMO regulators, MoA (Ministry of Agriculture) and MoS&T (Ministry of Science and Technology), are acting together to collude with Monsanto and the biotech Industry to willy-nilly impose GMOs into Indian agriculture, in the face of clear scientific evidence that suggests abundant caution. The latest Affidavit filed by the MoA with the MoEF as a co-signatory, but in a subordinate position, opposed the Report of the technical experts committee tooth and nail, and clearly provides this evidence.

After having the recommendation of the parliamentary standing committee on agriculture in its report in 2012, the recommendation of the technical expert committee constituted by the Supreme Court in its 2013 report, and in the light of the Government of India pushing and attempting

---

[92] *The Economic Times.* 2013. 'Put genetically-modified crop trials on hold for now: Supreme Court Panel'.

[93] EU follows a stricter and more formal regulatory system regarding GM food: Evenson and Raney. 2007. *The Political Economy of Genetically Modified Foods*, p. 294.

to go ahead with the approval of field trials of GM crops, it can be said that in India, the issue of field trials and commercial cultivation of GM crops is viewed with varied perspectives. On one hand, the government is insisting on the commercial use of GM crops; on the other hand, the parliament and the Supreme Court are taking a cautious approach. The executive is insisting on the field trials of GM crops, whereas the legislature and the judiciary are against it. In the Indian democracy, the organs of the state—the executive, the legislature, and the judiciary—ought to respect, consult, consider, cooperate, and coordinate with each other. The Constitution of India has given equal rights, obligations, and status to all the three organs of the state. When the two organs of the state are either opposing or being cautious about GM crops, whether the executive alone can proceed with approving field trials of GM crops is a big question that the constitutional experts will have to address and answer.

Government of India Decides to Promote GM Crops

The Government of India, under the aegis of Ministry of Agriculture and Farmers Welfare, has a different view regarding the GM crops which it presented in its affidavit before the Supreme Court. The ministry has argued that transgenic technology is absolutely needed for India's food security, which showed that the Government of India seems to have realized the importance of progress of science and research in this field. The Ministry of Agriculture and Farmers Welfare feels that without biotechnology and GM crops, the interests of Indian agriculture and food security will be jeopardized.[94] The affidavit was filed in response to the report of the technical committee appointed by the Supreme Court on 10 May 2012 in matters related to field trials of GM crops.[95] The affidavit states:[96]

---

[94] Ministry of Agriculture and Farmers Welfare has filed an affidavit before the Supreme Court in the case of *Aruna Rodrigues* v. *Union* of India, Writ Petition (Civil) No. 260 of 2005 and Writ Petition (Civil) No. 115 of 2004.

[95] See generally, http://www.gmwatch.org/index.php/news (last accessed 10 August 2014).

[96] 'Scientists tell Indian Government GM cannot deliver food security.'2013.Availableathttp://www.gmwatch.org/news/archive/2013/14639-scientists-tell-indian-government-gm-cannot-deliver-food-security (last accessed 25 June 2016).

The demand for food and processed commodities is increasing due to growing population and rising per capita income. There are projections that demand for food grains would increase from 192 million tons in 2000 to 345 million tons in 2030. Hence in the next 20 years, production of food grains needs to be increased at the rate of 5.5 million tones annual.

A real-time analysis of this scenario provides sufficient justification for strengthening, intensifying and introducing cutting edge science and technology for increasing crop productivity in India. With rising population, decreasing size of agricultural holding, reduced soil fertility, resource constraints in terms of land and water coupled with uncertainties arising out of agro-climatic conditions, the blend of genetic modified technology with other conventional tool is the valid solution for ensuring food security for its increasing population.

Furthermore, it was argued that the report and recommendation of the technical expert committee for the 10-year moratorium on field trials of GM crops would mean a complete stop to agri-biotech research application, and that it would result in indirectly benefiting MNCs, with India having to import GM technology from abroad. Echoing and furthering the approach of the Ministry of Agriculture and Farmers Welfare, the then prime minister of India, Manmohan Singh, stated in February 2014: "[W]hile safety must be ensured, we should not succumb to unscientific prejudices against BT crops." With one broad stroke, Singh had colored all opposition to the immediate approval for GM food crop trials as "unscientific."[97] The government seemed to be in complete support of GM crops and clearing of field trials of the same. It is believed by the government that to cater to the hunger needs of the country as well as to realize the dream of ensuring food security to the mass population of the country, GM crops is the only option at the moment. In terms of biosafety, India has a strong regulatory system; however, the existing system now requires ensuring that all GM crops undergo rigorous reviews and safety assessments prior to their import, field-testing, or release. What is lacking is an effective implementation mechanism.[98] The present government led by

[97] Available at http://www.business-standard.com/article/economy-policy/gm-crops-pm-revealed-his-assertive-self-to-push-for-trials-114041400028_1.html (last accessed 9 August 2014).

[98] Sreenivasulu and Chettery. 2016. 'Agricultural Biotechnology: Environmental Law and Other Policy Considerations'.

Narendra Damodardas Modi, also seems to be in favor of the field trials of GM crops.[99] The debate over the introduction of GM food crops in India has remained polarized. The GM crop industry, many scientists in the field, the union agricultural minister, the prime minister, and some states such as Andhra Pradesh, Gujarat, Punjab, and Rajasthan believe GM crops would rejuvenate agricultural productivity and lead to food security. The other side is packed with NGOs, most of the other state governments, some political parties, scientists, academicians, and green activists who remain skeptical. More importantly, they are wary of the unassessed and potentially irreversible damaging impact of releasing GM crops on people and environment. Nevertheless, at the moment, more than 100 GM food crop varieties are moving through the experimentation pipeline and can fundamentally change the nature of food crops and production in the country. These crops include rice, wheat, okra, onion, groundnut, bamboo, tomato, apple, cucumber, sugarcane, cabbage, cauliflower, tea, coffee, corn, ginger, ragi, yam, castor, sunflower, mustard, black pepper, pea, soybean, papaya, cardamom, carrot, banana, tobacco, orange, pearl millet, potato, and pulses. Very recently, in February 2014, the Government of India enacted the National Food Security Act[100] which promised food security and right to food to 70 percent of the people—given the fact that currently the population in India has gone beyond 120 crore , 70 percent of the same would be 84 crore. Without the promotion of GM crops, it may not be possible to realize the dream of ensuring food security to the mass of 84 crore people as visualized under the said act.

GM Crops: The Reality Check

There are certain aspects that one has to consider in the biotechnology industry, in terms of its effectiveness, implications, and performance. Probably, while formulating regulatory framework as well as in terms of progressing with biotechnology, trade, and agriculture, we need

[99] Available at http://www.gmwatch.org/index.php/news/archive/2014/15544-modi-govt-condemned-for-allowing-gmo-field-trials (last accessed 10 August 2014).
[100] National Food Security Act, 2013 was enacted by the Parliament in 2014.

to consider these aspects. The first commercialized crops came into being around 16 years ago, and to this day, only two commercially viable transgenic traits are present, which are cultivated mainly in three countries— the US, Brazil, and Argentina—that grow 77 percent of all GM crops. An overwhelming majority of countries worldwide do not grow GM crops. GM crops are grown on a mere 160 million hectare that comprise 3.2 percent of the global agriculture land.[101] Just four crops cover 99 percent of this area—soybean (47 percent), maize (32 percent), cotton (15 percent), and canola (5 percent). The two traits that have been commercialized the most are: (1) Bt crops for insect resistance with genes from the soil bacterium *Bacillus thuringiensis* inserted for a new toxin to be produced within the plant to kill insects and (2) HT or herbicide tolerance, where the engineered plant is able to withstand herbicide sprays of particular kinds. HT is the overwhelming trait in commercially grown GM crops today.[102]

Biotechnology Regulatory Authority of India Bill, 2013

Amidst all the controversies on biotechnology regulation, there was an attempt to provide for a regulatory framework for biotechnology in India in 2006, with the establishment of a national authority by the Government of India. In November 2006, the government, through prime minister's office, issued a directive to the Department of Biotechnology to act as a nodal agency for the establishment of national biotechnology regulatory authority. The directive intended to provide for a single-window agency for regulating all the issues pertinent to biotechnology and GE products. It was proposed that the existing mechanism in whatever form would continue to function until the national authority comes into power with full autonomy on the matters related to biotechnology. Initially, in 2008, National Biotechnology Regulatory

---

[101] *FAO Statistical Yearbook.* 2012. Macroeconomy (page 42) Computed from FAO Data: Total Agricultural Land is 4.9 Billion Hectares, available at http://www.fao.org/docrep/015/i2490e/i2490e01c.pdf (last accessed 3 March 2016).

[102] *ISAAA Brief 43-2011 Executive Summary*, February 2012, available at http://www.isaaa.org/resources/publications/briefs/43/executivesummary/default.asp (last accessed 3 March 2016).

328 LAW RELATING TO BIOTECHNOLOGY

Authority Bill[103] was introduced for the first time, which, in its original form, was lacking in several aspects. The bill was introduced to establish a regulatory body to ensure safe and responsible use of biotechnology and also to meet India's international obligations under Convention on Biological Diversity, 1992, and the Cartagena Protocol on Biosafety, 2003. The Ministry of Science and Technology, in its road map in 2007, has also recommended for the establishment of a regulatory authority for biotechnology. The bill, however, is plausibly not in accordance with many established principles.[104] A look at the content of the bill would give us an idea about what its framers intended to bring in:[105]

1. In accordance with the precautionary approach undertaken in Principle 15 of Rio Declaration, the Cartagena Protocol mentions potential risks associated with modern biotechnology and the need for adequate level of protection and, thus, the necessity of the following precautionary principle. The bill in its present form, however, does not seem to have adopted the precautionary principle.

2. The nodal agency sought to be established is such that it will be effectively controlled by the Department of Biotechnology whose objective is to propagate biotechnology. This leads to the possibility of bias and is against the law laid down by the Supreme Court in a number of cases that there should not be any likelihood of bias itself.[106]

[103] The Biotechnology Regulatory Authority Bill 2008 intends to set up a central regulatory authority for handling issues concerning biotechnology in India.

[104] Greenpeace India Society. 2012. *BRAI Bill: A Threat to Our Food and Farming: A Legal Assessment*, 9 February, available at http://www.greenpeace.org/india/en/publications/BRAI-bill-A-threat-to-our-food-and-farming/ (last accessed 25 June 2016).

[105] The Biotechnology Regulatory Authority of India Bill, 2013, available at http://www.prsindia.org/uploads/media/Biotech%20Regulatory/Biotechnology%20Regulatory%20Authority%20of%20India%20Bill.pdf (last accessed 25 June 2016).

[106] *Justice P.D. Dinakaran v. Hon'ble Judges Inquiry Committee and Others* (2011) 10 SCR 1064.

3. The role of state governments is curtailed under the bill by taking away their power to reject GMOs. This is against the principle of federalism.

4. The statutory committees sought to be set up under the bill do not include members from economics, public health, and social science background. Thus, it does not contain provisions to establish wholesome policies by not having participation from a larger section of society.

5. Given that there is still uncertainty over potential impacts of GM crops on human health and environment, it is perhaps essential to have a long-term independent biosafety assessment mechanism, which is currently missing.

6. The bill curtails participation of public in decision-making. This is against various principles and accepted norms of public participation such as those recognized under Article 23 of Cartagena Protocol and Principle 10 of the Rio Declaration.

7. The punishment for providing wrong and misleading information under the bill is only up to 3 months of imprisonment, which does not seem to be a strong enough deterrent. The key deterrent effects such as absolute liability and polluter pays principle, on the other hand, are also absent.

Index

About the Author

Sreenivasulu N.S. is professor of law at the West Bengal National University of Juridical Sciences (NUJS), Kolkata, India. He has been teaching and researching on intellectual property and technology law since 2003. He has contributed to the legal jurisprudence through his writings and has about 95 research publications. His recent book is *Law Relating to Intellectual Property* (2013). He got invited to Max Planck Institute for Innovation and Competition, Germany in 2010 and 2013 as Max Planck Fellow. He was also invited by Edinburgh Law School, University of Edinburgh, Scotland in 2010 as Script Fellow for academic visit. He was invited by Global Illuminators, Malaysia in 2014 and Madison County Bar Association, Madison, USA in 2016 as resource person/chair for academic events on intellectual property and biotechnology law. He is also associated with World Commission for Human Rights at London. He is an advisor/peer reviewer/editor for international journals such as *Journal of Intellectual Property Rights*,

California and *International Journal of Intellectual Property Law*, Turkey. He was offered Microsoft Chair Professorship by Microsoft India in 2014. He has provided consultation and advisory services on policy matters on IPR to the Government of India.